DRILLS, TAPS
& DIES

Other books by the same author

Hardening, Tempering and Heat Treatment
Milling Operations in the Lathe
Model Engineer's Handbook
Soldering and Brazing
Spring Design and Manufacture
Workholding in the Lathe
Workshop Drawing

Writing as T. D. Walshaw
The I.S.O. System of Units
Ornamental Turning

Drills, Taps and Dies

Tubal Cain

SPECIAL INTEREST MODEL BOOKS
Special Interest Model Books Ltd.
P.O. Box 327
Poole
Dorset
BH15 2RG

First published by Argus Books 1976

Reprinted 1989, 1990, 1993, 1996, 1999

This edition © 2002 Special Interest Model Books Ltd.

ISBN 0 85242 866 9

Phototypesetting by Photocomp Ltd., Birmingham
Printed and bound in Great Britain by Biddles Ltd, *www.biddles.co.uk*

Contents

Author's Preface

In preparing this book I have taken a somewhat different approach from previous volumes on the same subject. First, I have given a great deal more attention to drills. After all, we drill a score or so of holes for every one which is tapped! But there is another reason. The apparently simple twist drill is, in fact, a very complex tool indeed; even now its cutting action is not fully understood, but enough has been discovered in recent years to give a better appreciation, and to reveal facts which will enable us both to drill better holes and to extend the life of the tool. I hope too that what I have written will encourage readers to treat their twist-drills with the respect they deserve!

The second part of the book, dealing with the tapping operation, is also quite new, and is the result of work which I carried out some years ago. I cannot be the only one to have been struck by the fact that in Industry the main preoccupation is with tap *sharpening*, whereas the jobbing worker, the model engineer, and the amateur generally, is beset with problems of tap *breakage*! The analysis of the effects of depth of thread engagement on fastener strength, which I can only summarise in these pages, showed that the problem lay not with the tap, nor yet with any lack of skill on the part of the user, but rather with the choice of tapping drill. I hope that after reading this the choice of tapping drill will be made somewhat easier *and* that you will find that tap breakage is drastically reduced! However, this leads to one consequence; you will, I hope, soon realise that there is no "correct" tapping drill for any thread — you must make a choice, and this means that the tables of drills may appear to be somewhat more voluminous than you are accustomed to. I have tried to make this choice as easy as possible for you — for 90% of your work a single size will serve. But my own experience shows that for the other 10% — and it is almost always in this region that tap breakage occurs — a sensible use of the tables will improve matters for you considerably.

Finally, I have left out all consideration of screw-cutting in the lathe — this in spite of the fact that almost all the male threads I cut with a die are partially screwcut first. But there is another book by the same publishers dealing with this in detail, and I see no point in "vain repetition"! In any case, there are enough problems in the use of drills, taps and dies without extending the book to deal with lathe-work as well.

I hope that you will derive as much enjoyment from reading this book as I have in the writing of it — and that my good friends in the tap and die industry will not be too upset at seeing a drop in their sales!

T.D.W   Westmorland, July 1986

SECTION 1

TWIST DRILLS

Types of Drill

Many thousands of years ago some Bronze Age warrior found that he could drill holes in timber by using his spear with a twisting motion. In later times a similar tool was made specifically as a drill, and this type is still used in woodturning. Fig. 1 shows some, made by Holtzapffel about 150 years ago and still in service. The edge is formed as in Fig. 2a and as the blade is tapered in thickness a fairly sharp point results. These are intended for use in a lathe, so that the direction of rotation is uniform and there is a zero rake cutting face and a clearance on the opposite face. If used as a hand tool the shape would be as in Fig. 2b, with what is in effect a negative rake on both edges. These tools are very effective in wood, especially that of Fig. 2a in hardwood. The point gives a good start, and used with the hand-rest on the lathe the directional control is good, the tool cuts freely, and quite deep holes can be drilled provided the chips are cleared frequently.

When used for metal, however, this type is too weak. The point soon collapses, and the curved profile causes

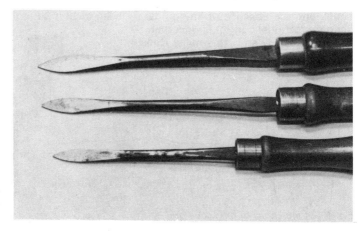

Fig. 1 *Spear-point drills for use in the lathe.*

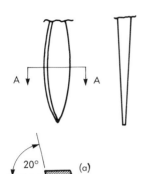

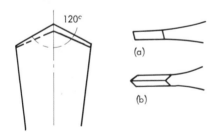

SECTION AA

difficulties. The spear point was, as a result, developed until the **Spade** form evolved – Fig. 3. The profile is shown in Fig. 4, and again, this may be arranged for unidirectional rotation as at (a) or for reversing use at (b). The latter is found in the very small sizes used by watchmakers, who drive the drill with a bow and bit-holder. Fig. 4a is still to be found in some blacksmiths' shops, as they are very easy to make, and I show one in use in Fig. 5. If carefully made, with equal lips, hole size is fairly well controlled, but due to the relatively weak shank and the absence of any guiding action directional control is uncertain – this is especially so with smaller drills, say below about 10mm (⅜in.) dia-

Left, **Fig. 2** *The spear point (a) For unidirectional rotation. (b) For use as a hand-tool.*

Above, **Fig. 3** *A home-made spade drill.*

meter. However, the spade drill is still used today at the two extremes of size – very small and very large – and we shall deal with these later.

To be sure of drilling to size and keeping the hole straight the **fluted drill** was developed during the 19th century. This, Fig. 6, was, in effect, a cylindrical piece of tool-steel (hence the term "drill rod") with two flutes milled along part of the length. The cutting edges were formed by grinding a cone with a small relief angle – known as the "relieved cone" – so that the actual cutting edge was very similar to that of Fig. 4a. The drill is much stiffer than the spade type and has the great advantage that the cylindrical part guides the drill as it

Fig. 4 *Cutting edges of the spade drill.*

Fig. 5 *A spade drill in use.*

Fig. 6 *The straight fluted drill.*

penetrates. There is space for chips within the flutes, but not enough to hold all the swarf, so that for holes more than about 1½ diameters deep it is necessary to withdraw periodically to clear the chips. The defect of the design is that the centre of the drill has a web, so that the actual point has a very unfavourable cutting angle – Fig. 7. We shall be looking at this aspect of the cutting action later; suffice to say at this stage that apart from the very point, which acts like a punch, these edges do *cut*, even though the rake angle is extremely negative. The main parts of the cutting edges offer a nominally zero rake. Later refinements of the straight flute drill led to the cylindrical surface being relieved, to reduce friction, and to a tapered web, so that the stiffness was maintained with a thinner web at the actual point.

Woodworkers had, of course, used auger-type drills for a very long time. These, with their helical grooves or flutes, "wound out" the chips and allowed very deep holes to be drilled quite easily. Such drills could not (at that time) be used for metal, as they were not stiff enough either to accept the much higher thrust or to resist the much greater torque applied compared with that in woodwork. They were, in effect, a flat piece of tool-steel twisted into a helix. However, it was not long after the introduction of the straight-flute drill that machinery was developed to permit this flute to be cut (or forged and then twisted) in helical form, and the present day "twist drill" was born. The point form is exactly the same as that of the straight flute drill; the helix angle provides positive rake, and the edges of the flutes are again relieved,

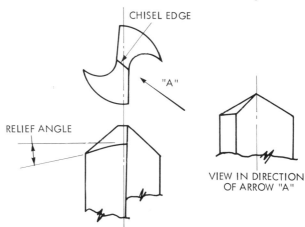

Fig. 7 *The point of a straight-flute drill.*

9

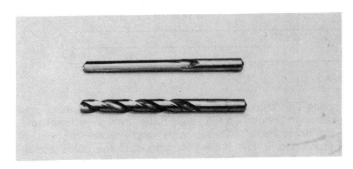

Fig. 8 *Comparison between straight-flute and twist drills.*

so that the guiding action is almost as good. Compared with the straight-flute drill, however, the twist drill is much less stiff, and more prone to wander when drilling deep holes. Fig. 8 shows the comparison between the two types. The twist drill is now almost the universal tool, but those who still have straight flute drills available will use these in preference to twist drills for brass, gunmetal, and similar materials. As we shall see later, a developed form of the spade drill, with a special, very stiff, shank, is used for large diameter holes. Twist drills themselves now come in a

wide variety of types, but before considering these it will be as well to look at the cutting action, so that the reason for so many can be understood.

The Drill Point

There is far more to a drill point than meets the eye! First, to avoid confusion later, let us "name the parts". Fig. 9 shows the parts of the drill as a whole, the upper diagram being a taper shank and the lower a straight shank drill. The recess shown on the latter is usually found only on drills above 6 mm or ¼ in. diameter, and not always then. Most

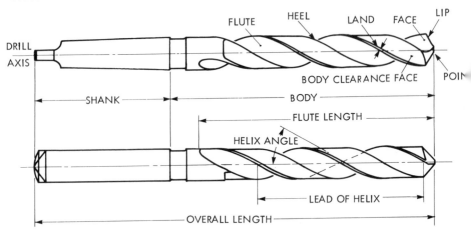

Fig. 9 *Naming the parts of a twist drill.*

Courtesy B.S.I.

10

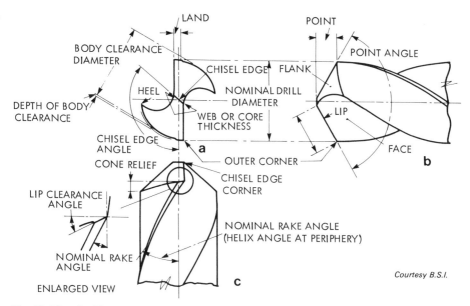

Fig. 10 The significant elements of the point of a twist-drill.

of the names are self-descriptive; the "heel" is the tail end of the helix, and the "land" is the narrow part of the body which bears on the sides of the hole to give guidance. The "lip" is the cutting edge. The "lead" is the same thing as the pitch of a screw thread, and results in a helix angle. Here is the first important thing to notice. The *lead* is the same right across the drill section, but as the diameter is reduced as we get to the axis of the drill, so the *helix angle will diminish*. If there were no web between the flutes this angle would become zero at the centre.

Fig. 10 refers specifically to the point of the drill. At (a) we have the end view. If the point were a simple cone it would present an actual point at the centre. But the two halves of the conical part are given a *relief* – see 10c – which serves the same function as the clear-ance angle of a lathe tool. The effect of this at the centre of the web is that instead of forming a conical point we find a *chisel edge* which is like the roof of a house – there is *no* "point". Have a look at a largish drill – a new one, or one which has been properly ground. Note that the line of this chisel edge lies at an angle to the line across the two lips or cutting edges. This is the *chisel edge angle*, and it is important that this be correct. Fig. 10a also shows the usual shape of the flutes, and you can see how easily a straight-flute drill can be made, given a suitable milling cutter. At 10b is shown the all-important *point angle*, the maintenance of which pre-sents difficulties for those without proper sharpening jigs. It is VITAL that this point angle be exactly equally disposed either side of the drill axis. Similarly, the two *lip lengths* must be equal if oversize

11

holes are to be avoided. (These matters will be gone into again when we come to drill sharpening.) Note also the spot marked *outer corner*. This is the most vulnerable part of the drill point and the spot where wear first occurs in normal use.

Now look at Fig. 10c. Notice how the conical relief provides a clearance behind the cutting edge or lip, and that the helix angle provides the rake to the cutting edge. When these figures are quoted they refer to the angles at the outer corner (see 10b) — as already observed, the rake angle diminishes along the length of the lip towards the drill axis. Note that the rake angle is marked "nominal". This is because the *effective* rake angle is different — a matter we shall look at later.

Now, there is no need to sit up all night "learning the names of the parts"! But it will help to have these diagrams to refer to later. In the meantime, looking at Figs. 9 and 10 you will see that many of the dimensions — or "parameters", to use the 'in' word — are fixed by the maker. There is nothing you can do to alter the helix angle (though you can alter its effects) or the flute depths. But as soon as you start using the drill you will wear the chisel edge, the outer corner, the lip and, after extensive use, the drill diameter across the lands. When you resharpen the drill, whatever means are used, you have the power to

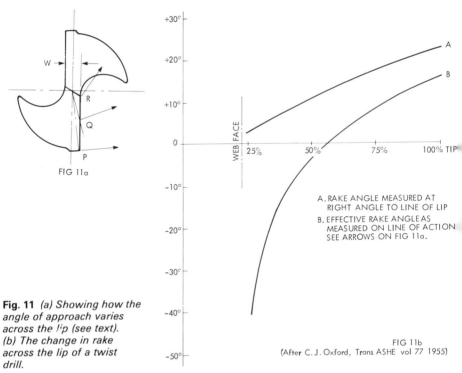

A. RAKE ANGLE MEASURED AT RIGHT ANGLE TO LINE OF LIP

B. EFFECTIVE RAKE ANGLE AS MEASURED ON LINE OF ACTION SEE ARROWS ON FIG 11a.

FIG 11a

Fig. 11 (a) Showing how the angle of approach varies across the lip (see text). (b) The change in rake across the lip of a twist drill.

FIG 11b
(After C. J. Oxford, Trans ASHE vol 77 1955)

alter the point angle, the cone relief (and hence the lip clearance), the chisel edge angle and the lip length. A lot of variables. It is surprising how well drills put up with them all!

The Cutting Action

The action of the drill edge is often likened to that of a knife-tool in lathe work, but the apparent similarity is misleading. Look at Fig. 11a which shows the end view of the point. We have already noticed that the rake angle formed by the twist in the drill will diminish towards the centre – from P to R on the drawing. However, this view shows that the angle at which the edge meets the work also changes across the length of the lip. This lip is displaced from the centre of rotation of the drill by half the web thickness, w/2, so that while at P the approach is very nearly at right angles to the lip, at R it is about 40°. This means that the rake angle will be reduced still further – in fact, at R the effective rake is negative, the transition occurring (with the normal-flute helix) at Q, about half-way between P and R. Fig. 11b shows this effect, which is exactly the same as that found when a boring tool is set above centre-height on the lathe.

On the chisel edge, however, the approach of the edge remains almost constant – provided, that is, that the drill is properly ground. The rake here is about 56° *negative*, which appears to be an almost impossible cutting situation. However, it must be remembered that when drilling the axial forces are VERY much greater than those in turning*, and provided that this axial force is maintained this apparently ineffective

---

*See Appendix 'A', page 81.

edge will actually form a chip. At the very centre of rotation the point acts like a rotating punch. A contributory factor to the cutting effectiveness of this part of the drill point is that the heat produced in this negative rake area softens the workpiece slightly. On the other hand, with work-hardening materials like some classes of stainless steel the slightest relaxation of feed pressure will harden up the workpiece locally and in many cases result in destruction of the chisel edge of the drill.

Fig. 12 shows some remarkable illustrations of the chip formation, which are sketched from photographs made during research by Mr W.A. Haggerty of the Cincinnati Machine Tool Co. in 1961. These were, in fact, micrographs, obtained by photographing sections of the material after an almost instantaneous stop to the action of the drilling rig. In each case the direction of cut is from left to right, and it will be seen that even as close as 0·010 in. to the centre of rotation of the drill there is some semblance of a "chip" being formed. The chisel edge does *cut* – provided the axial force on the drill is sufficient. The main problem in this region is the limited space through which the chips can *escape*. They are often found to be "wiped" out of the cutting region, simply because they have been forced through a very slim aperture.

The relevance of these factors to the drill user is that the sharpness of the chisel edge is just as important as that of the lips. Further, if the chisel edge angle (Fig. 10) is not correct then the cutting action will be impaired; the drill forces will be increased and hence the local heating, with the end result that the chisel edge may be destroyed completely. The integrity of the chisel edge of the drill is as important as that of the

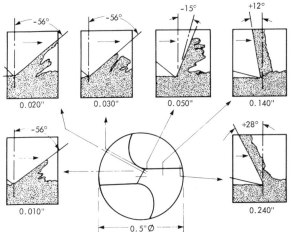

Fig. 12 Sketches taken from micrographs of the chip formation at various points on the cutting edge of a drill. (After W. A. Haggerty)

remainder of the point, and becomes of paramount importance when drilling work-hardening materials. A final point to note is that the surface of the cone which provides the *clearance* for the lip – the main cutting edge – is in fact the *rake* surface of the chisel edge, over which the chips must flow. The finish of this area is often neglected, as users feel that there is "obviously" no need for a good finish on a surface which apparently is not in contact, but careful dressing of the wheel before grinding – or even a touch from an oilstone on larger drills – can improve the performance of the chisel edge quite markedly.

Other aspects of point geometry

The point angle A change in the point angle – normally 118° – does more than make the drill more or less "pointed". It affects the shape of the lip as well. See Fig. 13. Drill designers arrange the shape of the flute so that at the "general purpose" angle of 118° the lip or cutting edge is a straight line. Reducing the included angle as at (b) produces a backwards curving lip, and increasing it gives one which curves forwards – (c). For almost all materials the straight lip gives the best results, but there may be occasions when a lot of work has to be done on either very soft or very hard

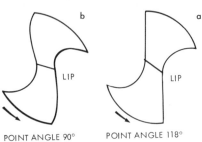

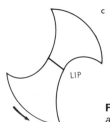

Fig. 13 Effect of point angle on lip shape for a normal helix drill.

material when an alteration of the point angle may be justified. However, the clearance angle (see Fig. 10) also has an effect, and we will consider both of these angles in this respect later.

More important than the actual angle is its symmetry. It will make very little difference if the total included angle is 115° or 125°, but if one lip lies at 57° to the drill axis and the other at 61° – still providing 118° total – the effect can be serious – even a difference of one degree will have an effect. See Fig. 14a. The flatter lip 'C' is doing all the cutting and will, as a result, suffer excessive wear. Further, the thrust is all concentrated on this lip, tending to drive the point over sideways. An oversize hole is probable and if the difference in angle is more than slight there is a risk that the drill will run off course. The situation is exaggerated if the lips are of unequal *length* as well – almost unavoidable if the point angle is askew. As shown in Fig. 14b, the hole is definitely oversize. The corner 'D' cuts away the metal left by the short lip 'E', with consequent risk of this corner failing prematurely, and the hole will be oversize, almost cer-

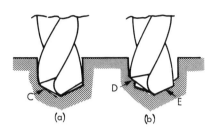

Fig. 14 *(a) Effect of unequal angles. (b) Lips of unequal length.*

tainly run off line, and will have a very rough finish.

Clearance Angle Fig. 15. At first sight it might seem that provided there *is* a clearance all would be well, the more so as the clearance measured at the periphery of the drill increases towards the centre. However, feed rates in drills can be quite high and if the clearance angle is insufficient the depth of feed during partial rotation of the drill could exceed the dimension (a) and the drill would rub. In addition, the chip cut by the chisel edge, shown shaded at (b), has no route of escape except along the

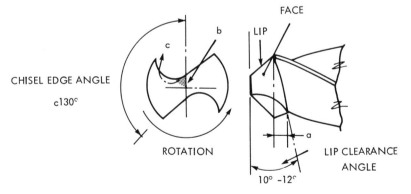

Fig. 15 *Clearance angle. The arrow (c) shows the path of the chips leaving the chisel cutting edge at "b". See text.*

15

Fig. 16 *Compound drill point for drilling cast iron.*

arrow (c). If the lip clearance angle is too small then these chips will be wedged, will probably weld to the drill face, and the point will be damaged. For general purposes the clearance angle, measured at the periphery of the drill, should be from 10° to 12° – though again, it may be desirable to alter this for some materials. The chisel edge angle shown in Fig. 15 should be about 130°, and this angle is a good indicator that the drill has been correctly ground.

Choice of angles

The standard point will meet all normal work, and it is hardly worth altering it unless a considerable number of holes

Fig. 17 *Brad-point twist drill for timber or soft sheet metal.*

is to be drilled in an awkward material. However, the following are the usually recommended figures.

Material	Point angle	Clearance angle
Mild Steel	118°	10°–12°
Tough Steel	130°	12°
Brass, Bronze	118°	15°
Al. & alloys	100°	15°
Copper	100°	15°
Plastics	90°	10°
Medium Cast Iron	118°	10°–12°

For softer grey cast iron and malleable iron a *compound point* is often used. See Fig. 16. The outer 25-30% of the lip is ground to an included angle of 90°, keeping the same clearance angle as for the standard drill. This provides the strong chisel edge of the standard grind, but gives more favourable cutting action on the lips. Some researchers recommend a full 90° point as for plastics, but the compound point is favoured because standard drills are easily altered.

The application in recent years of twist drills used in electric hand tools for drilling timber has resulted in the development of a point with a flat end – Fig. 17. This "Brad Point" is ground with about 15° clearance angle and the chisel edge formed into a projecting bradawl-like centre; such drills have very thin webs at the point, tapering to a larger thickness at the shank. This type of drill has found applications in drilling soft sheet metal, but woodworking drills have soft temper and should not be used.

The Helix Angle

We have seen that the helix angle determines the cutting rake. This angle also provides the "auger" effect which helps to bring the chips out of the hole and also curls the chips so that they take up as little room as possible. (The flute volume is considerably less than that of the metal removed.) As always, a compromise is necessary between the various requirements.

The **Straight-flute drill** offers zero rake and no auger effect. Chips must be cleared by "woodpecker" action – repeated withdrawal of the drill. But the lips are very strong and the body of the drill is much stiffer than that of the twist drill. Such drills (Fig. 6) are rather difficult to find amongst tool-dealers these days, but they are the ideal for brass.

Next comes the **Slow helix drill** with a helix angle of about 22½°. Fig. 18. This provides some auger effect without unduly weakening the lip compared with the straight-flute type. In "production" work on the copper alloys, brass especially, such slow helix drills are manufactured with larger flute spaces than normal, but for "jobbing" work the normal section is quite satisfactory.

The **Normal helix** is about 40°, an angle which has been found by experience to cover almost all the jobbing worker's requirements. The lip is not quite as strong as that of the straight-flute or slow helix, but the axial pressure is less and less power is needed to drive the drill. The faster helix gives much better chip clearance and, on steel especially, a better "curl" to the chip.

Quick helix drills with an angle of about 45-50° were originally developed for use with the light alloys, and provide much better chip clearance, especially on deep holes. The weaker lip is of less importance with these soft materials. They are also used on copper, though drills ordered specifically *for* this metal

will have a slightly slower helix. In almost all cases, quick helix drills will normally be supplied with wide flutes and smaller lands on the body than on the standard drill. Quick helix drills are also used even with steel for drilling the long oil-holes in engine crankshafts, though such are usually provided with a thicker web than standard and always have a rather "special" point.

For *general use* the reader is advised to stick to the standard helix and standard point on drills until "trouble" makes a change imperative. There is, however, one exception to this. The received wisdom when drilling brass is to stone the lip to reduce the rake to zero. There is a tendency when drilling brass and

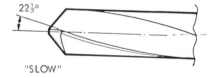

"SLOW"

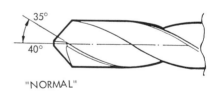

"NORMAL"

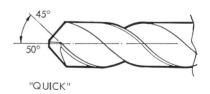

"QUICK"

Fig. 18 *Helix angles normally available.*

17

similar alloys for the drill to "walk into" the workpiece – indeed, with brass it has been known for the chuck to be pulled off its taper when the drill breaks through. The reason for this is complex, but it will be appreciated that these metals are relatively soft and, more important, soften up even more when hot – as at the chisel edge of a drill. The downward resistance to the feed at this point is therefore much reduced. The shear force is quite high, and if you look at the point of the "normal" drill on Fig. 18 you will appreciate that as the chip passes over the lip there will be a component of force acting *downwards*. This can, in some circumstances, exceed the axial resistance to feed, and the drill then "walks into" the metal.

To prevent this by stoning a flat on the drill lip is quite satisfactory – it is not really necessary to have a very wide flat, either. But the after-effect is that you will either have to regrind the drill, or buy another, for use on normal materials. Regrinding the drill takes time, needs a proper drill-grinding jig, and unless you are well practised in drill sharpening can result in a faulty point. Further, we have already noted that the web of the drill gets thicker as you approach the shank, so that in due course the chisel edge gets wider and wider. If, on the other hand, you meet the case by buying a second drill for normal materials, I suggest that, instead, you consider investing in a slow-spiral (or even a straight-flute, if you can find one) reserved for copper alloys only. They are more expensive, but will last a long time.

To sum up, the standard helix angle will serve for almost all materials, and even for the exceptions will give satisfactory results if care is taken. In production work the SLOW helix is used for Brass, Beryllium-copper, Manganese steels, hard Plastics and for stone-drilling. The QUICK helix would be used for Aluminium and its alloys, Nimonics, Titanium alloys, Magnesium alloy and in the endgrain of hardwoods. For all other materials the normal helix would be employed.

Drill Lengths Fig. 19.

The usual straight shank drill is the **Jobber's**, shown at (b), and derives its name from the fact that it was used in "jobbing" workshops, where frequent changes of drill size were needed. (In production work a taper shank drill is used, avoiding the need for a chuck.) These drills are designed for holes from about 4 to 10 diameters deep. The majority of holes – especially tapped holes – are less than this, and the *Stub* drills, (a), designed for holes from 2 to 4 diameters deep, have a number of advantages. They are much stiffer and in many cases do not need any "start" in the way of a centre-punch or preliminary drill point. I use nothing else for tapping-drills. For deeper holes the *Long* series is used, and will cope with depths of from 5 to 12 diameters. There is available an *Extra long* series, for holes up to 30 diameters deep, but these are seldom needed!

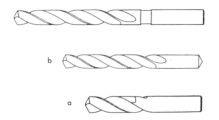

Fig. 19 *"Long", "Jobber's" and "Stub" length drills. "Extra long" are also available.*

There is little price difference between Jobber's and Stub drills – the latter are very slightly more expensive in the sizes up to ½in. – but the Long type are about 2½ times the price of Jobber's and Extra Long are very expensive indeed! In cases where extra length is needed to get at an awkward spot it is almost always possible to drill the end of a piece of mild steel rod (about 1½ to twice the drill size in depth) and soft solder the drill in place. See Fig. 21. Soft solder will be quite adequate up to about ³⁄₁₆in. dia. if the drill shank penetrates about 2 drill diameters; above about ⁵⁄₁₆in. I use low temperature brazing alloy (BSAg1 or Ag2). This is quite safe on a high-speed steel drill. For extensions where the hole itself is deep it is possible, with care, to braze the drill onto a shank *smaller* than the hole diameter – Fig. 22. The shanks of H.S.S. drills are relatively soft and can be drilled or machined reasonably well, using a low cutting speed and plenty of cutting oil. However, in using an extension drill of this type it is imperative that frequent withdrawals be made to clear chips. The drill will certainly wander off course otherwise.

It is, of course, possible to transform broken jobber's drills into stub type, but I would again emphasise that the *web*

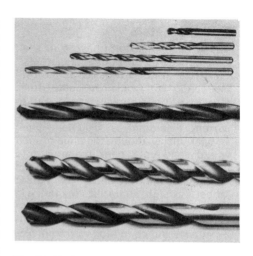

Fig. 20 *Various lengths and helix angles of twist drills.*

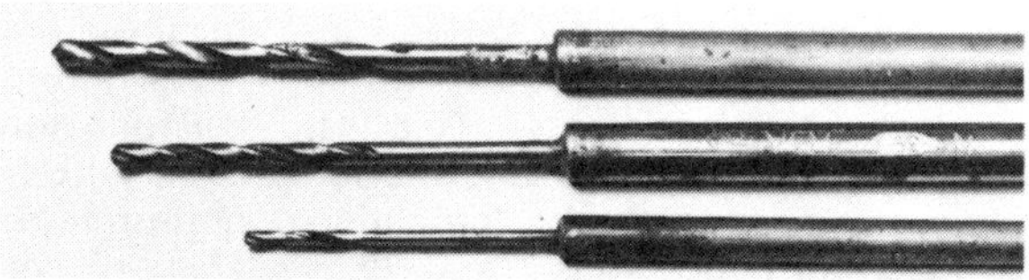

increases in thickness towards the shank. This means that a conventionally ground drill point will have a chisel edge of excessive length. In converting such broken drills it is advisable to use one of the methods shown on page 34 – either resorting to point-thinning or using the four-facet method of grinding (see page 34). For use in the lathe, however, where a combination centre drill ("Slocombe Drill") is used to start the hole, this may not be necessary.

Fig. 21 *Drills lengthened to reach an awkward place, soldered into extension-pieces.*

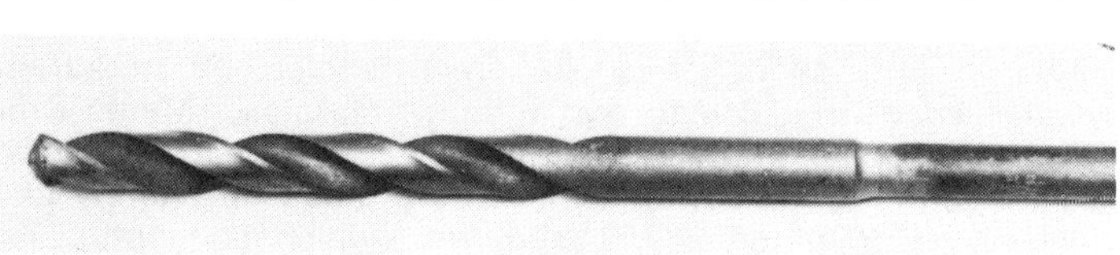

Fig. 22 *Drill lengthened to make a deep hole. It is brazed to an extension slightly smaller in diameter.*

Drill Diameters

The largest twist drill I have ever seen was between 5in. and 6in. diameter, but for such holes today a Spade Drill (see page 26) would be used. Stock size twist drills nowadays go up to about 4in. The smallest drill listed in a current catalogue is 0·0001in. (yes; 1/10,000 inch!), though the price of such would send your bank manager into fits – even more so for the special drilling machine needed! In regular jobber's straight shank lists the range is from 0·20mm (about 0·008in) up to 20mm, and for taper shank types from 2·4mm (³/₃₂in.) up to 4in., with from No. 1 to No. 6 Morse tapers. "Micro Drills" can be had ex-stock from 0·05mm dia. up to 1·45mm dia. in steps of 0·05mm at quite reasonable prices; these have a uniform shank diameter of 1mm up to 0·80mm, and 1·5mm shanks from there upwards. My current drill catalogue is a shade over one inch thick, and the ordinary jobber's drill section has eight double-column pages. There is a drill for every size hole!

Years ago there were three families of drills; *Imperial*, starting at ¹/₆₄in. and rising by ¹/₆₄-in. steps upward, *Metric*, starting at about 0·025mm and in steps of perhaps 0·01mm in the smaller sizes, to 0·25mm for the larger drills, and the Morse *"Number and Letter"* series, running from No. 80 at 0·0134in. up to letter Z at 0·413in. diameter. This last series was very much used, though there were disadvantages; the difference between (e.g.) No. 77 and 76 is 0·005in., but between 76 and 75 only 0·002in. This series was declared *obsolescent* nearly 20 years ago at time of writing, and is now discontinued. True, if you order a "number drill" you will be supplied, though it will be metric – e.g. 3·26mm for No. 30 – and it will cost you almost twice as much as the recommended alternative of 3·25mm – 0·01mm or 0·00039in. smaller. So, today there are two ranges of drills only which comply with BS 328; the Imperial, or "inch" sizes, and the millimetre range. "Number and Letter" drills are still supplied in U.S.A., but even there they are mostly I.S.O. metric equivalents. The Inch range works to intervals of ¹/₆₄in., though decimal sizes can be supplied if the quantity is large enough. The metric range provides very close intervals indeed at the smaller end, there being about 600 sizes in the jobber's range between 0·20mm and 20mm.

Preferred sizes

In order to rationalise this very wide choice both the British Standards Institution and the International Standards Organisation (I.S.O.) have listed a range of drills which considerable consultation suggested would meet all normal requirements. However, it must be emphasised that these preferred sizes are directed to the *machinery designer*, in an attempt to limit the number of different sizes he calls for on the drawing. The *user* is in a different position, for if the drawing calls for (e.g.) "7mm ream" normal workshop practice *may* call for a non-preferred size. Similarly, it is very likely that tapping sizes will be non-preferred drills. All sizes are available, of course, but the normal tool dealer may stock only the B.S.I. list. It should be noted, however, that the classical "Metric Drill Set", 1 to 6mm (or 1 to 10mm) in steps of 0·1mm is made up entirely of "preferred" sizes, which is helpful. **Table II** shows the preferred sizes for both Imperial and Metric drills for straight-shank types only, but covers all lengths from stub to extra-long. The difference in price between these and intermediate sizes depends on demand – it can be as much as 30% or as little as 10% – sometimes nil, for a "preferred Imperial" drill is a non-preferred size in the metric range and vice-versa! (Table I gives metric conversion.)

Number and Letter Equivalents

We have to work to many old drawings on which all drill sizes are given by number or letter, many older practitioners still use them even on new drawings and, of course, most users still have stocks of such drills, with the need to replace the odd one now and again. Table III shows both the *recommended* alternative and the *"exact"* (within 0·005mm or so) equivalent, together with the decimal sizes in inches. You will see that the B.S.I. recommendations are seldom more than 0·01mm, or, at worst, 0·02mm, (0·00039 to 0·00078in.) different from the "exact" figure. The difference is larger with the larger sizes, but even there not sufficient to be significant. In a few cases the same metric size is equated to two adjacent numbers: e.g. No. 59 and 58 – but the difference between the drill numbers is also small.

Users in U.S.A., where "Number" drills are still quoted, can use the table to interpret metric sizes in terms of their standard. For others, I suggest that the range 1mm to 6mm × 0·1mm will serve our needs just as well as the old system. After all, it is no great matter if the drill is 0·114in. diameter instead of 0·113in. (No. 33) or 0·116in. (No. 32) – even if the drill cuts to size exactly!

Drilling speeds

The recommended drilling speeds found in Production Engineering reference books are designed to achieve the most economical production rate, taking into account the cost of sharpening and replacing the drills, and allowing for the fact that the drills will be sharpened by skilled toolmakers on proper machines. On the other hand, they also assume that the drill is in use continuously. For the jobbing worker and model engineer quite different considerations apply – chiefly the wish to *avoid* having to resharpen for as long as possible! In addition, almost all commercial drilling is done under power downfeed, and this makes a great deal of difference. For manual drilling (or, as we used to call it in the works, "pin-drilling"!) the feed control is very variable and in selecting the speed a great deal does depend on your own feelings about "feeding". Indeed, a very good rule is that the speed should be selected so that both you and the drill "seem to be comfortable", with reasonable chips emerging steadily from the drill. There is no EXACT, single, "correct" speed for any type of drill; even a change in the cutting oil used will make a difference. So, use your judgement, and don't hesitate to depart from "the book" if you find that you get better results.

Having said that, you do need something to give a guide as to where to start, and to that end I show overleaf a set of speeds which I have found will give reasonable results, when using the lubricants suggested on page 76. Use these as a guide.

It is often suggested that very small drills – below, say, 2mm – should be run "as fast as you can". This is true when faced with the ordinary drilling

Drill dia., mm. in.	2 0·08	4 0·16	6 0·24	8 0·32	10 0·39	12 0·47
Material						
Aluminium, Dural, Tufnol	—	—	4000	3100	2300	1900
Brass, Freecutting M.S.	—	3400	2500	1900	1450	1250
Bronze, Grey C.I., BDMS, Gunmetal, Phos. Bronze	4600	2500	1700	1260	1000	800
Mall. Iron, Monel metal, Silver steel, Stainless steel	3600	1700	1150	880	660	550
Hard cast-iron	1800	1000	650	500	400	330

Drill dia., mm. in.	14 0·55	16 0·63	18 0·71	20 0·79	22 0·87	24 0·98
Material						
Aluminium, Dural, Tufnol	1600	1400	1250	1070	1000	900
Brass, Freecutting M.S.	1050	900	800	700	640	560
Bronze, Grey C.I., BDMS, Gunmetal, Phos. Bronze	700	600	540	480	420	400
Mall. Iron, Monel metal, Silver steel, Stainless steel	480	420	400	380	350	200
Hard cast-iron	280	240	220	200	180	160

If the machine has a limited number of speeds, choose the next LOWEST to that given in the table.

machine, with a top speed around 3000 rpm, but there are machines which can run much faster. Here the use of "as fast as possible" can, in fact, blunt or even break small drills. It has to be remembered that tiny drills are very likely to bend a trifle under the axial load, even though that may be very small. The drill then starts to "whirl" and this can set up enough stress to break it.

The makers of one brand of special HSS Micro-precision drills suggest that the highest speed should be used at about 1 mm dia., and then be sharply reduced BELOW this figure. The table below gives a few examples, the speed being in thousands of rpm.

Drill dia., mm	0·2	0·4	0·6	0·8	1·0	1·2	1·4	1·6	1·8
Aluminium	6·5	12·5	16·0	18·5	20·0	20·0	19·5	18·5	17·5
Brass	6·0	10·7	14·0	15·5	16·0	15·8	15·2	14·4	13·2
GM, Bronze	2·5	5·5	8·0	9·7	10·3	11·0	10·7	10·0	9·2
Steel (EN8)	4·0	5·8	8·0	8·3	8·8	8·8	8·0	7·8	7·2
C.I.	2·1	3·6	5·1	6·0	6·5	6·2	5·8	5·4	

This table serves to emphasise that the tip cutting speed used in conventional machining is not relevant to the drilling operation. With the smaller drills especially – 3 mm and below – it pays to run a little slower when drilling holes more than 3 diameters deep.

At the other extreme, lack of power prevents the use of "proper" speeds with drills of 10-12 mm upwards. If a reasonable feed rate is maintained our machines just do not develop enough power. It is better to drop the speed and keep the feed-rate up, so that reasonable chips are formed. The worst thing to do to a drill is to allow it to rub – remember, the chisel edge can only cut if the axial force is large enough.

DRILLS OF OTHER KINDS

Rather an odd title for a section heading, but I can think of no other which fits! There are literally hundreds of "other kinds" of drill, mostly for special purposes, but some made in quantity. However, I propose to limit myself to those most likely to be used by model engineers, amateur mechanics, and the small jobbing shops.

The Combination Centre Drill

This is sometimes called a "Slocombe" drill, Slocombe being the original producers. See Fig. 23. Type 'A' is that commonly used. As the name implies, its purpose is the formation of centre-holes in the ends of workpieces for use in the lathe between centres. The former way of producing such centres was first to drill a pilot hole and then follow with a 60° countersink. The invention of the "combination drill" has saved billions of man-hours! (Earlier still they used the "square centre"; I still have one, but very seldom use it.) The combination drill is also used as a starter for twist drills in the lathe, and sometimes in the drilling machine also, but it is not a *necessary* adjunct, just a useful one. Provided that the twist drill is smaller in diameter than the diameter of the

tapered centre-hole this procedure gives a good start, but care must be taken to avoid drill chatter when starting if the reverse is the case.

Type 'B' is a refinement. The second 120° taper forms a recess which will protect the centre-hole proper. This type should be used on workpieces where the centre-hole will be needed during its service life. Type 'C' is a very refined centre-drill. It produces a hole in which the lathe centre bears with line contact and, in general, gives a more accurate "run" to the workpiece. Further, the friction is less in service and there is a reduced risk of overheating the lathe centre. On the other hand, if the tail-stock exerts too much pressure the poppet may have to be adjusted more frequently as the wear will be faster. It is really for use on light precision work, and I use this type only on my Lorch precision lathe – and use no other type on it.

Table IV (page 88) gives the dimensions of the current standard drills – though 'C' is designated as "Type R" in BS and ISO standards. Note that the best designation to use is the *body* diameter, though the pilot diameter is often used. I have given the approximate pilot diameter in inches as well as in millimetres. Note that type 'A' can be had single-ended, but the dimensions are exactly the same. I have included in the table the old, obsolete, 1950 "Imperial" sizes as well. These should be defined by the BS No. – BS1, BS5, etc. Incidentally, these drills can be had in body diameters up to 25mm, but few of us have chucks that size!

When used to produce lathe centre-holes the penetration for types 'A' and 'C' should be such that the diameter of the tapered hole is between 75% and 85% of the body diameter. With type 'B' the guard recess should be 85-90% of the body diameter. The drill should be discarded if the pilot length, 'l', is ground so much that it falls below 75% of the original length, otherwise there is risk of the centre-point bearing on the bottom of the pilot hole. This cannot happen with type 'C'.

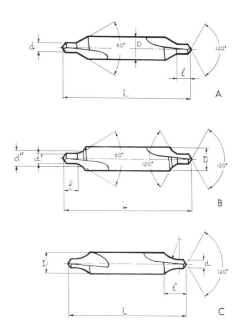

Fig. 23 *The "Slocombe" or Combination centre drill. (a) The normal type. (b) Type producing a guard recess. (c) Precision type producing a curved profile to the centre hole.*

The Spade-bit drill

We have already noticed this one, in Figs. 4 and 5, page 8, but a few words may not be amiss. There are two types – those you can make yourself, and the larger commercial type. I use the former not infrequently, usually when I want a drill a bit larger than my chuck will hold, and smaller than one of the three Morse taper shank drills that I own. Some are fairly crude, but nowadays I use ½in. silver steel. The end is heated to yellow red (1000°C) and flattened. Take care – don't go on hammering if the metal cools below bright red. The width is thus increased and the section thinned. Rather than forge heavily if the drill is to be only just over ½in. I would probably go down to ⅜in. stock, but experience shows that the larger the shank the better. The end is cleaned up and, if

necessary, reheated and straightened. I then turn it to the shape needed in the lathe, using a point angle of 120°, the diameter about 1/16in. oversize. After cutting to length the tool is annealed at 760°C – "blood-red" – and then finally shaped, leaving just a shade on for grinding. Harden from 790°C (cherry red) and temper as required for the material on which it is to be used. The cutting edges are ground as shown in Fig. 24. For use on steel etc. I form a few nicks as seen at 'b'; these break up the chips and make the work easier for the drill; note that the nicks in the two lips must be at different radius.

25

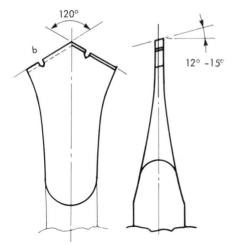

Fig. 24 *Point form of a home-made spade drill.*

The *commercial* type of spade drill is a different kettle of fish. See Fig. 25. Used for holes of 1 in. diameter upwards in steel, cast iron, bronzes and brass, especially the more difficult types, but *not* for the stickier aluminium alloys, which may form false edges to the lips. There is, of course, no auger effect, so that "woodpecker" action is essential, sometimes with fluid washout, but increasingly this problem is being got over by inverting the drilling operation so that the chips *fall* out – a procedure I

first saw adopted in my place of work about 50 years ago, when drilling long holes in large connecting rods.

The main feature of these drills – apart from the spade itself – is the very stiff holding bar. This, coupled with the narrow cylindrical land, helps to prevent the drill from wandering. In fact, they produce very straight holes, with a good finish. However, they MUST be used with copious lubrication (indeed, I use force-feed even on my Fig. 24 type). The bits are interchangeable, though, of course, a different arbor would be used for a 5 in. cutter compared with a 2 in. Carbide tips are available. A variant of this type has a flat cutting end for flat-bottoming holes.

On the face of it it might seem that home-made drills of this type would be quite practicable, but there are a few snags. First, I could not imagine that anyone could *drill* a (say) 2 in. hole in a 3½ in. centre-lathe – or even in a drilling machine with but a ¾HP motor! Second, it is *vital* that the spade be accurately centred on the arbor, otherwise the drill will hole oversize. However, no doubt a "poor man's version" could be devised!

The Watchmaker's "Foret"

This is, in effect, a tiny spade drill of the type shown in Fig. 4b, page 8. In the

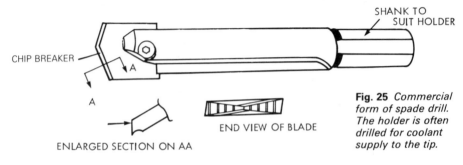

CHIP BREAKER

SHANK TO SUIT HOLDER

A

A

ENLARGED SECTION ON AA

END VIEW OF BLADE

Fig. 25 *Commercial form of spade drill. The holder is often drilled for coolant supply to the tip.*

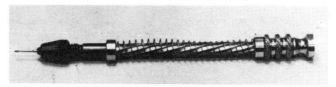

Fig. 25a *Watchmaker's "Foret" – a miniature spade drill, shown fitted in the typical archimedean drill-stock.*

old days these were used with a drill-stock and bow, giving a back and forth rotation, but nowadys a small archimedean drill-stock is employed. Fig. 25a. Indeed, apart from the flatter point angle and harder material they are very similar to the "Fretwork Drills" once very common. They can be had right down to 0·1 mm dia., and are numbered in "tenths" – e.g. a No. 12 will be 1·2mm dia. and a No. 5½ is 0·55mm. The shanks are normally uniformly 2m dia. for use in collets, but the smaller ones may be found with 1mm shanks. To cater for the increasing use of small motor-driven bench drills some makers now supply these little drills ground for unidirectional rotation; if used in a bow or archimedean drill-stock these cut in one direction only, but as they cut faster, no time is lost.

These little fellows are very hard and fast-cutting, in brass especially, and have a very fine point. I use them a lot, not so much for drilling as for *preparing* to drill. Once the position of the hole is marked out, the point of the foret is passed along one of the marked-out lines until is felt to engage with the cross-line. The drill-stock is then held upright and a few strokes will make a sizeable spot-hole. I use a No. 10 drill

(1mm) but anything between No. 6 and No. 15 will do. This "pop" is quite enough to start small drills – up to about 4mm or so – and provides a more accurate "spot" for the subsequent use of a centre-punch when that is needed.

They are, of course, drills in their own right, and provided the hole is less than about 4 diameters deep there is no problem with chip clearance.

The Core Drill

These drills are designed for the opening out of smaller holes, especially those cored out in castings or rough punched in plate. The two main features (Fig. 26) are that they have a much thicker web and hence are much stiffer than a twist drill, and that they have three, rather than two, flutes. They have no chisel edge and *cannot* be used to drill a hole from scratch. There must be an existing hole which is larger than the diameter across the web. The reason for using three flutes can, in part, be seen from Fig. 27, which exaggerates the effect to make it clearer. If an ordinary twist drill is used to enlarge a hole by more than a very small amount a "lobed" hole is almost certain to be formed. On initial contact one lip will, unless a jig-bush is used and centring is

Fig. 26 *The 3-flute core-drill compared with a standard twist drill.*

27

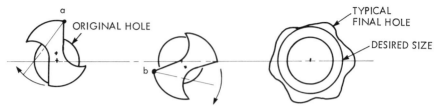

Fig. 27 *The "lobing" action of a normal twist-drill when enlarging a hole.*

absolutely precise, make a deeper contact than the other. A twist drill is very flexible so that the point will tend to wander, and the second lip rotate about the first, as at 'a'. This second lip then digs in, and the drill point rotates about *it*, at 'b', causing the other to dig in about ⅓ rev. later. This process continues, ('c') as it is self-sustaining, and a "lobular" hole results. A test bar of the drill diameter may "fit" the hole but only "where it touches" – there will be gaps all round.

The three-flute drill avoids this risk, first, because it is very much stiffer, and the point tends to "stay put". Second, there are no "opposite flutes", so that the risk of chatter is reduced. In addition, the point geometry is slightly different, a change made possible by the fact that the drill does not have to remove its full diameter of metal.

These drills are usually made with taper shanks; there is a straight shank version, but NOT for use in a chuck; it is provided with a small tang, and is

devised for use with special collets. The use of a chuck "adds to the lack of rigidity". Core drills are available from about 8 mm (⁵⁄₁₆ in.) dia. up to 50 mm (2 in.) and are sold in "reaming sizes" – from 0·2 to 0·4 mm below "preferred sizes" – as well as in the standard diameters. However, used alone they produce a hole finish considerably better than will a normal twist drill. As the name implies, they can be used very successfully to open up rough cored cast holes. In the absence of a proper core-drill, a stub drill will be less likely to cause lobing, being much stiffer than a jobber's pattern. If the latter is used, then each step should remove only a very small amount of metal.

The D-bit

This type of drill is really a guided boring bar – Fig. 28. The body is very slightly less in diameter than the desired hole – a "running fit" to it – and is flattened on one side so that there is just over half a circle remaining. The

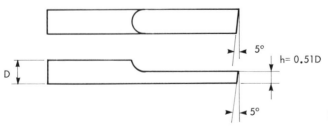

5°

h= 0.51D

5°

Fig. 28 *The D-bit.*

D

Fig. 29 *The half-round drill.*

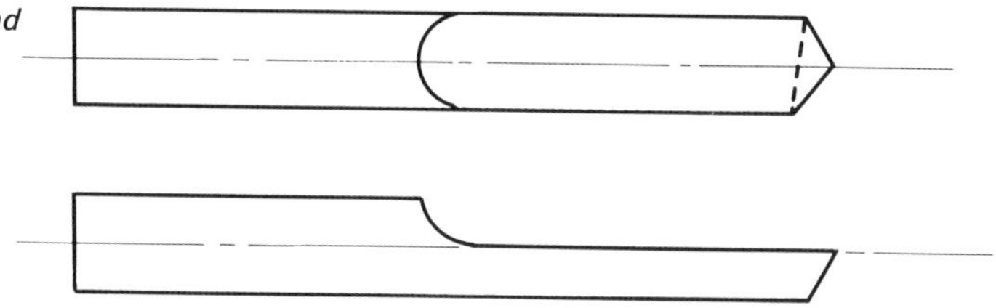

drill cuts only on its front edge, which has relief as shown. Provided that the finish on the body is good such drills will work to reamer size and finish, but they MUST be started by drilling a hole of the required size, of depth equal to around half the diameter. These drills are usually made up from silver steel, but the ground finish is not always good enough and if used direct from the rod, the grinding marks should be polished out. I prefer to machine down from larger stock, polish to give perhaps 0·0002in. clearance in the hole, and then file or machine the flat. It is usual to make the dimension 'h'=0·51 D – i.e. about D/100 more than the half-diameter. Both the top and front faces should be got up with an oilstone, taking care to keep both quite flat. The bit is sharpened only on the front face. Cutting speed is best somewhat below that of a twist drill of the same diameter. The feed-rate should be kept up, and to ensure a good finish it is wise NOT to

"woodpecker"; the chips usually clear well enough. If the hole must be deep – above 5 diameters – then it is best to drill in stages, using D-bits both times. The first can be woodpeckered, but the second will take full depth, as the chips will be much narrower. D-bits with a very small cross-clearance, say ½°, can be used to flat-bottom holes if a dead flat face is not essential.

The Half-round drill

This type seems to be almost unknown. It is, in effect, a D-bit with a point, Fig. 29. They are commercially available from 0·35mm (about No. 80) up to 12mm and ½in., both metric and imperial. They will, with care, drill to 10 diameters deep, leave a reamer finish, and unlike the D-bit need no start hole. Woodpecker operation may spoil the finish and in production work an air-jet is used on sizes above about 6mm to remove chips. They are not normally used on steel or cast iron. For best

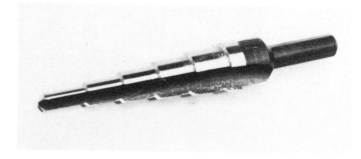

Fig. 30 *A stepped "Conecut" drill for use in thin material. If has but a single flute.*

Fig. 31 *The Conecut drill in action.*

results cutting speeds should be about 25% less than for normal drills and feed-rates about 50% more. They can, of course, be made from carbon steel stock (silver steel or drill rod), but it is more important that the point be truly central and that the half-cone be properly relieved with clearance. Bought commercially they cost about four times the price of a jobber's twist drill, and are available from about 0.5mm up to 10mm dia.

The "Conecut" drill

This is a relatively recent introduction. It is, in effect, a single-flute drill, and it is designed for cutting clean holes in sheet metal. They can be had in a plain conical shape, typically running from 3-14mm, 6-20mm, 16-30mm dia., designed for opening out existing holes, or in the form seen in my illustration, with the diameters in fixed steps, which can start from a centre-pop, again with a range of maximum and minimum diameters. They are very effective indeed. That shown in Fig. 31 (a "UNIBIT") has six steps, cutting from ³⁄₁₆in. dia. to ½in. dia. in steps of ¹⁄₁₆in. It is drilling tough steel sheet, and the limit of thickness is about ⅛in. in mild steel. The Unibit is available in various sizes up to 34mm (1³⁄₈in.) maximum diameter, but those with maxima above about 18mm (¾in.) need a pilot hole slightly smaller than the smallest step; the usual way of dealing with this is to use the next size down Unibit and to enlarge with the larger one.

SECTION 3

DRILL SHARPENING

From what has gone before it is evident that sharpening a drill means more than just putting an edge on it. See Fig. 32. The two lips or cutting edges 'a' must be of the same length, and the point angles 'b' must both be correct and equally disposed about the drill axis. The clearance 'c' must be sufficient. The leading edges of the lips must be level as shown at 'd' – this will follow automatically if 'a' and 'b' are correct. And the chisel edge angle must be correct, at 'e'. (The figures shown are those for a 118° point with 10-12° clearance angle, normal helix.) To grind a drill by hand calls for considerable skill and experience, and this can only come from practice. You must expect to "get things wrong" for quite a while – or at least, "get things not quite right"! – but it will help if you practise without a grinding wheel to start with. Seek out the largest unused drill you have – mine is $^{31}/_{64}$ in., a hole size I seem never to have used. Find a vertical smooth metal surface and cover this with marking blue. Offer the drill point and practise the motion needed evenly to coat the drill point

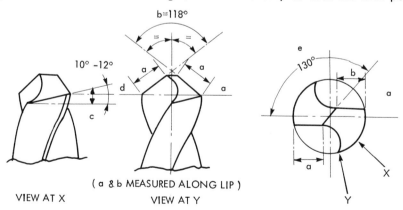

(a & b MEASURED ALONG LIP)

VIEW AT X   VIEW AT Y

Fig. 32 *The critical elements needing attention when sharpening a twist-drill.*

with blue. You will find that you have to (a) hold the drill at the correct angle; (b) gently raise and lower the left hand (the one holding the point end); (c) move the right hand slightly up and down and side to side at the same time; and (d) slightly rotate the drill with the right hand – all these motions occurring simultaneously!

Practise this until you can get a reasonable cover of blue on the point. Then take an old drill and try it on the grindstone. Note that you must use the flat side of the wheel, not the periphery. Yes, I know that "everyone says" that you should never do this, the reason being that wheels are not very strong sideways. But the force used in drill-grinding is and should be *very* slight and I can assure you that no harm will come. The ideal wheel is, of course, a cup wheel, but few bench grinders will accept one. After a while you will find that you can get a reasonable approach

to the correct angle and clearance. You must now practise further to get both sides even. This will take you quite a while, but persevere! It is extremely doubtful that you will ever get a drill absolutely correct with off-hand grinding like this, but you should not have a *great* deal of trouble in achieving a tolerable point – so long as you practise. Remember, the aim is to take off the minimum amount necessary to renew the cutting edges of lip and chisel point. I always finish my drills with a medium fine India oilstone – you will remember that the clearance faces to the lips form the rake faces to the chisel edge, and should, therefore, be smooth.

Far better than off-hand grinding is to use a drill grinding jig. Fig. 33 shows the well-known "Potts", marketed since Mr Potts' death by Woking Precision Models Ltd. (There are others of similar design.) It is a reduced version of the man-sized jigs formerly used in industry, before

Fig. 33 *The Potts drill sharpening fixture for use on the bench grinder.*

special-purpose drill-grinding machines (costing thousands of pounds) became universal, but has none of the complications of vernier angle settings, micrometer feed devices etc. found on such. Properly used it will give a point very near to perfection *provided that it is set up and used correctly*. Follow the assembly instructions exactly, so that it is presented to the grinding wheel in precisely the fashion intended by the designer. The main problem in use is in the accurate setting of the drill point to the stop in the jig. This must be done very carefully. It would be a waste of my time to go through the instructions as, although I have a Potts which I have used for nearly 30 years there are other similar devices available and each has its own setting method. I think I should add that when properly set up these jigs often "look wrong". Don't be alarmed – just check through the instructions and make sure that you have not made a mistake, and then press on!

"Other Devices"

There are, of course, many devices other than these. Some are basically identical to the Potts and not to be disdained. After all, Mr Potts based his design on the more elaborate industrial type. Others often appear to be crude, but do serve their purpose – the restoring of the edge of drills used for "Do-it-yourself" projects etc. The handyman is seldom concerned with accuracy to the last hundredth of a millimetre or with drill performance, and for this class of work it is far more important that the edge should be keen than that it should conform exactly with the design geometry. Such cheap gadgets are much better than unskilled off-hand sharpening! Indeed, I confess that I have one, made entirely of plastic, which puts an

edge on by rubbing the point on a sheet of silicon carbide paper, maintaining the angles by the peculiar motion of its "wheels". It works, and with care can work quite well enough for me to prefer it to off-hand grinding when I am in too much of a hurry to set up my Potts or the Quorn. But it is used *only* on the drills I keep for use in the hand or electric portable drill-drivers. In which connection I would urge readers NOT to use their "workshop" drills, and especially not their tapping drills, for D.I.Y. odd jobs. Small drills are easily bent, and larger ones can suffer badly from both chisel edge and land wear when used with portable drilling engines. However, even "second best" drills should be kept sharp, and a cheap drill sharpener, used with care, is well worthwhile.

The "Four-facet" point

This type of point grinding has become increasingly popular following the introduction of numerically controlled machine tools, (though it has been in use since about 1905), as it is more or less self-starting and needs no preliminary centre-hole. The difficulty for the amateur user is that it is almost impossible to produce the shape by off-hand grinding, and an error in the profile is much more serious than in the case of the relieved cone point which we have considered so far.

Fig. 34 shows the arrangement. There is a double clearance angle; the primary angle, 'A', gives the major clearance and corresponds to that produced in a relieved cone drill. The secondary angle, 'B', is the cutting edge clearance. This secondary clearance is carried right to the centre of the drill, so that there is a *point* on the chisel edge instead of a flat; I have shown an end view of this

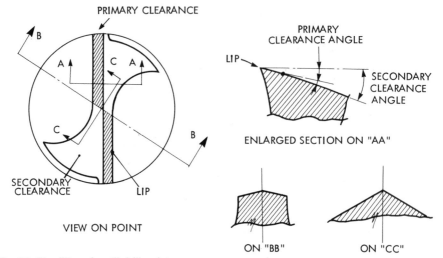

PRIMARY CLEARANCE

PRIMARY CLEARANCE ANGLE

SECONDARY CLEARANCE ANGLE

LIP

SECONDARY CLEARANCE

LIP

VIEW ON POINT

ENLARGED SECTION ON "AA"

ON "BB"   ON "CC"

ENLARGED SECTION AT DRILL POINT

Fig. 34 *The "Four-facet" drill point.*

type of grind in Fig. 34, and you can see the difference. If used with stub drills this grind completely obviates the need even for centre-punching (provided that you can position both drill and work sufficiently accurately) and on jobber's length drills a marked reduction in the tendency for the drill to wander from (e.g.) the centre-pop will be observed. There is, in addition, evidence that in deep hole drilling the drill runs much straighter.

So far there appears to be no simple drill-grinding jig on the market which can produce this grind, and the machines used in industry cost rather a lot of money! However, the Quorn tool and cutter grinder *can* grind the four-facet point (but can't grind the relieved cone without modification) and full instructions are given in Prof. Chaddock's book "The Quorn Tool & Cutter Grinder", from Argus Books Ltd. All the considerations about equality of lip length, accur-

acy of angle-setting for both point and clearance, and so on, apply equally to the four-facet and relieved cone points. I would emphasise this again: neither the use of a drill-grinding machine, a "Potts" jig, nor a "Quorn" will guarantee a good point to a drill unless the instructions are carefully followed and the drill properly set up in the holder.

Point thinning

On larger drills (say above 12-15mm dia.) other considerations apply, notably the need for strength to resist the high torsional and thrust forces. It is doubtful whether the four-facet grind offers any advantage in these larger sizes. However, to provide the necessary strength the web of the drill is quite large – perhaps 3mm (⅛in.) on a 15mm (⅝in.) drill – and the consequently wide chisel edge can cause problems both in starting and when drilling. To overcome this the point is usually "thinned" as shown

**The 4-facet point is *not* recommended for use on brass as it may cause "grabbing". See pp 17/18.

Fig. 35 *Reducing the chisel edge by point-thinning the web.*

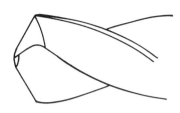

in Fig. 35. Two narrow grooves are ground along the line of the chisel edge, and this reduces it almost to a point. At the same time the acutely negative rake angle is reduced, thus providing better cutting conditions. On smaller drills – say below 10 mm (⅜ in.) – little advantage is gained, and the grinding operation involved becomes tricky. However, for drills between about 8 mm and 12 mm the increase in web thickness as the drill becomes shorter due to regrinding can become a nuisance, and in these cases point thinning may have to be resorted to when about half the usable length has been ground away. Point thinning is always used on drills above 12 mm (½ in.) when shortened in this way.

I have found considerable benefit when using large (above 12 mm, or ½ in.) drills in the lathe, the reduction in thrust making the difference between a "struggle" and almost normal drilling. But with anything much smaller there is little benefit and below 10 mm I never thin the point, even on well-shortened drills. Thinning is a very delicate operation, and if not done exactly right the last state is worse than the first. It is important that the "nicks" cut in the chisel edge should be identical in length and preferable that they be the same depth also, otherwise the drill will cut large and probably run off line as well. There is also a risk of accidentally touching the lip cutting edge. A thin saucer-type grinding wheel is needed,

of course – the edge of a normal grinding wheel, even if a sharp corner, has too great an angle. To do the job properly needs a grinding jig, and Fig. 36 shows a set-up designed and built by

Fig. 36 *The Bradley point thinning fixture. (Courtesy Ian Bradley Esq.).*

35

Mr Ian Bradley. The wheel is about 100 mm (4 in.) dia. and as you can see has been dressed to a very thin edge. The drill can be set accurately in the holder and clamped up, after which the depth of the thinning groove can be adjusted and, finally, fixed when the holder is rocked towards the wheel. No doubt those who have a Quorn could devise a similar arrangement. I would NOT, however, advise any but the most skilful reader to attempt to thin the point of any drill less than about 12 mm (½ in.) dia. working freehand on the bench grinder. It is worth noting, when considering point thinning of a much-used drill, that the diameter across the lands is tapered, being largest (to the nominal diameter) at the point. This taper is typically 0·02 to 0·08 mm in 100 mm (0·001 to 0·004 in. in 4 in.) length This is to prevent binding in the hole. Clearly, a worn and shortened drill will be undersize on the diameter.

Land Wear

This last fact leads to another. The lands on the flutes (see Fig. 9) guide the drill, and also establish the diameter of the hole which will be cut – assuming that you have ground the point correctly! Even in ideal circumstances these lands will wear, but as soon as the point geometry is less than perfect there is considerable extra load applied to one or other of the lands. This increases the wear rate. Sooner or later the lands at the point end will wear down to a smaller diameter than that further up the drill. Then, when drilling a hole deeper than usual these unworn parts will rub on the sides of the hole and, in the worst case, actually jam the drill in the hole. *There is no cure*, other than shortening the drill back to beyond the worn land faces. So, if you find that a drill tends to jam it is worth while showing it to your micrometer to see if the land wear is excessive.

Drill Storage

From what has gone before it is evident that the bad old days when drills were jumbled together in a tin box are, or should be, abandoned to the mists of memory! Quite apart from the inconvenience when trying to find the right size I hope that what I have written will have shown that a drill is a complex cutting tool; the integrity of the cutting edges and the finish of the relief faces deserve our best care and attention. To this end, for many years die-cast stands have been available in which the drills stand up like soldiers, each in its correct hole. In the best of these the holes were sized so that no drill will fit into the next smaller hole (the small "number" drills being an exception) but were easy to remove when needed. This is an excellent method, the only objection being that the tools are not protected from dust. After a time oily dust collects in the holes and these have to be cleaned out, but I have kept most of my drills in this fashion for 40 years. Such stands can, of course, be home-made, using boxwood or some similar timber which does not absorb moisture. (Avoid mahogany like the plague; this wood can take up moisture sufficiently to rust the drills solidly into the block in time). Fig. 37 shows my range of tapping drills (all of "stub" length) in their box, which has a turned lid to keep them clean. A little care may be needed to ensure that there is a slight clearance in the holes, and on the smallest ones I enlarge them slightly with a taper broach.

More recently drill cases of metal with a cover lid have become available, which are fitted with a sort of display

Fig. 37 *A home-made box of stub-drills.*

Fig. 38, below. *Set of drills 1mm – 6mm × 0·1mm in metal storage case.*

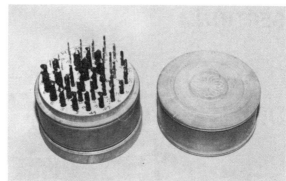

rack inside which opens up to present the drills in a convenient way. These are a bit more expensive but, over time, save their cost in keeping drills free from dirt (which can cause corrosion) and easily accessible. They have the added advantage to those who have a large range of drills in that the boxes can be stacked. Fig. 38 shows a set of metric drills 1 to 6mm × 0·1mm. Naturally, it is important always to put the drill back when finished with – this is no more than normal workshop hygiene – otherwise you have empty boxes or blocks and loose drills all over the place.

Spare or unused drills are best wrapped in oiled paper and stored in wooden or card boxes, though I have used strips of corrugated packing-paper, soaked in oil, and rolled the drills up within the corrugations. For the very smallest sizes – 1mm and below (Nos. 60 to 80) – I use fairly stiff paper, 80 to 100gm typing paper, oil-soaked, in which the drills are set like pins or needles, and then rolled up. Write the sizes on the paper before oiling it, though! One solution to the problem of small drills which I have seen was to keep them in glass or plastic pill-tubes, point upwards, with a cork stopper. The points can come to no

harm, and a drop of rust-preventative oil will preserve them. (Shell "ENSIS" grade 254 is excellent, as it possesses water-repellent properties.) My few very large taper shank drills I just put away in a rack, taking care that they don't bump anything, but wrap them if I know I am not going to use them for a week or so.

SECTION 4

DRILL CHUCKS

It is no use having a perfectly formed drill point if the drill itself is not held both firmly and true. In "Production" work, as I have said, a Morse taper shank is almost always used, though for smaller sizes in special machines straight shank drills may be held in collets. Even here special drills may be used, having a straight shank terminating in a tang – two flats at the end of the

shank. The collet directs the drill truly, the drive being transmitted through the two flats so that there is no risk of slipping. (On the Morse taper, however, the tang is there to ease drill extraction, and plays no part in the driving.)

The majority of readers will, however, use drill chucks, both in the lathe and in the drilling machine. I cannot emphasise too strongly the need for a good quality chuck. The duty is onerous in the extreme – the thrust on even a ¼-in. drill *can* be 100 lbf (50 Kgf) or so and if you work out the torque you will find that this puts a considerable tangential load on the chuck jaws as well. If the grip is not adequate then the drill must slip. Damage to the drill shank may put it out of true and will almost certainly set up burrs, and, of course, wear on drill chuck jaws is just as serious as wear on the jaws of a lathe chuck.

There have been many forms of chuck devised over the years, but the majority of users now employ the well-known JACOBS type with key-tightening of the

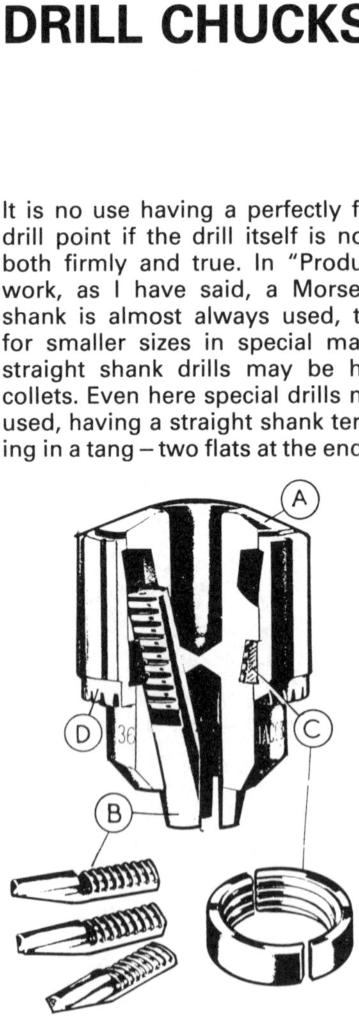

Fig. 39 *The Jacobs drill-chuck. (A) Body. (B) Jaws. (C) Split threaded ring, force fit into (D) Adjusting sleeve. (Courtesy Jacobs Mfg. Co. Ltd.)*

jaws, or the modern keyless types of which the ALBRECHT is typical. (The keyless types used on hand-drills are quite different, by the way, and though they are quite adequate for their purpose, they are not suitable for serious drilling.)

The Jacobs type is shown in section in Fig. 39. It comprises a body, 'a', machined to accept three jaws, 'b'. These are circular in section with helical teeth cut in the upper end, the lower end being formed into the gripping jaws. The teeth engage with a split ring, 'c', which is a tight press-fit to the adjusting sleeve 'd'. The inner face of the split ring is cut with threads matching the teeth on the jaws, and both the ring and jaws are hardened. When the sleeve 'd' is turned by the chuck key (which engages with the teeth on the lower edge) the jaws are forced down or retracted as required. The combination of the wedging action of the sloping jaws and the magnification produced by the screw action of the split ring provides a very strong grip indeed on the drill shank. The chuck shown in Fig. 39 is arranged for fitting to a taper arbor (the taper is an international standard) but they can also be had with a threaded socket – mainly for use on portable drilling machines. There are, of course, other makers than Jacobs, but the principle is the same except for details.

There is a variation in the design – the so-called "Super-chuck" which has a ballrace fitted between the split ring and the body, to take the reaction when the chuck is tightened. This means that for a given torque on the chuck key the jaws can exert considerably more force on the drill shank. Indeed, for smaller drills hand-tightening may suffice. They are, of course, more expensive, but well worth considering by those who expect many and frequent changes of drill size.

Keyless Chucks

Fig. 40 shows the Albrecht design of keyless chuck. The jaws are of tee section at the back face, and are guided in tee-shape grooves. Their movement is controlled by a thrust-plate working against the inner end, and this is operated by rotation of the screwed sleeve seen in the illustration. The thrust is carried on a ball-bearing. The chuck exerts enough force on the jaws to enable drills to be held securely by hand alone, no key being required, even on the largest size which accepts drills up to 16mm (⅝in.) dia. The smallest

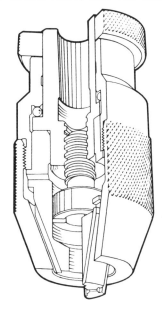

Fig. 40 *The Albrecht keyless chuck. The design enables even large drills to be gripped firmly when hand-tightened. (Courtesy Jacobs Mfg. Co. Ltd.)*

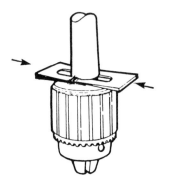

Fig. 41 *Removing a chuck from its arbor using a pair of wedges.*

ranges are noted for their precision, and the "0 to 3mm" does mean that it will hold a drill of almost zero diameter! The very smallest has a calibrated sleeve so that the chuck can be set to the drill diameter before loading; there can be few who have not struggled with a ½mm (No. 76) drill – somehow they always want to go into the spaces *between* jaws instead of in the centre! With this calibrated Albrecht there is nowhere for the drill to go *except* in the proper place!

As to chuck capacity, my largest Jacobs will accept 12mm (½in.) drills at the upper end, and I have no difficulty holding down to 1mm (No. 60) – it is now fairly old, and could hold even smaller when new. Albrecht type chucks are expensive, so that I use only the small 0-3mm size (this is not the very smallest) but wish I had one going up to 6mm (¼in.) for use in the Lorch lathe! It does depend a great deal on the class of work done in the shop; those concerned with small scale models, or clock and instrument work, would not need anything much larger than the Jacobs No. 32, taking up to 10mm (⅜in.) and the very small Albrecht, but the usual type

of pedestal drill with ½HP motor will cope with 12mm (½in.) and the Jacobs No. 34 is appropriate; either can be used in the lathe tailstock when fitted with the appropriate arbor.

Chuck Troubles

Like any other mechanical device chucks are subject to wear, and in due course some remedial action will be necessary. The first problem is: *removing the chuck from the arbor.* The Jacobs taper (which is an international standard for chuck fitting) is, like the Morse, a "wedging" taper and after some years in place the arbor may be extremely difficult to remove. The solution is simple. Look at Fig. 39 again. You will see that both at the bottom of the taper socket in the body 'a' and behind the jaws there is a recess with a "drill point" end. The metal is quite soft here, and a hole can be drilled through so that the arbor can be removed by using a brass or copper drift. However, a better way is to drill *and tap* this hole, so that a forcing screw can be used. Chucks up to 8mm capacity (⁵⁄₁₆in.) may be tapped 6mm or ¼in. BSF. From 8 to 12mm capacity (up to ½in.) M10 or ⅜in. BSF is suitable. The problem for many is that they have no other chuck in which to hold the drill! But all is not lost. Set the chuck in the tailstock, jaws fully retracted, and grip the drill in the lathe 3-jaw self-centring chuck; then drill in the normal way. I now drill and tap chucks when new, having learnt the lesson some 30 years ago. In this case it is best to hold the chuck by the body – away from the chuck key holes – in the lathe 3-jaw and drill down the taper hole. This way there is less risk of chips getting into the jaw-ways. Even so the chuck must be washed out thoroughly afterwards.

Fig. 42 *Dismantling and re-assembling a Jacobs chuck using a stepped driver-ring in the bench vice.*

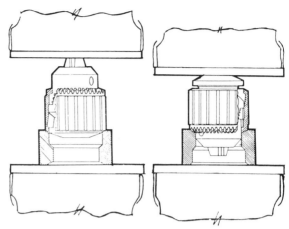

Where the chuck is mounted on an arbor greater than No. 2 Morse there will be a shoulder on the arbor. In this case a pair of thin "folding wedges" can be used (see Fig. 41) which can be squeezed together using the vice – or a hammer if you must. This is the only way possible for the Albrecht type of chuck, as the internal construction does not allow the drilling of a centre hole.

Jaw Wear

The jaws tend to wear most at the point, so that in time the drill is held only at the end of the shank. This not only allows the drill to run out of true at the the point but can also cause slippage unless excessive force is used on the chuck key. Damage to the drill shank results and, of course, the jaws wear even more. Contrary to supposition, dismantling of these chucks is quite practicable and the fitting of new jaws from time to time is normal practice. In the case of the Albrecht type it is advisable to send for the leaflet of dismantling instructions, and I will say no more than that no special tools are needed. The Jacobs keyed chucks do

need either a vice with about 4 inch opening or a mandrel press. Look again at Fig. 39. The ring 'c' is a force fit in the sleeve 'd', but only over its own width. As soon as the sleeve has moved about 10mm the ring is slack and all more or less falls apart. To dismantle you must first make a ring – and I suggest a stepped ring as shown in Fig. 42, which will serve for reassembling also. After adjusting the sleeve so that the jaws are about half-extended set the chuck up on the smaller diameter of the ring and apply pressure on the jaws. On chucks of any age the pressure required will be large, and it helps if the vice is tightened and then the movable jaw given a good bump with a rawhide mallet. As soon as the ring has started to move, the rest of the travel is fairly easy.

The faces of the jaws *can* be trimmed with a medium India slipstone until the bearing face is of uniform width full length. This is "better than nothing" but treated this way the jaws will no longer hold drills at the smallest limit of the chuck's range, and there is no certainty that all three jaws are evenly worn or stoned. New jaws are relatively cheap

41

and it is better to replace them. A replacement split ring will come with the new jaws. In re-assembling – after cleaning the inside of body and sleeve, and re-greasing – it is vital that the jaws be fitted in their correct guides. In some cases the jaws are numbered, in others they have one, two or three little grooves set in the recess at the end of the thread. The split ring should be greased and then assembled over the jaws – it will only go on with the two halves correctly mated – and then the sleeve can be pressed on again, using the forcing ring as before. This job is not difficult, and full detailed instructions (the above is just an outline) come with the spare parts. And to allay any unnecessary worry – those chuck jaws which I have HAD to replace have all seen 15 to 20 years' service!

Chuck-drop

This is the term I have invented for the situation where the chuck comes free from the arbor whilst drilling. As remarked earlier, it can happen when drilling brass at a high feed-rate just when the drill breaks through if precautions have not been taken. But it can, occasionally, become an epidemic. There can be two causes. On a new arbor – or even a new drilling machine – it can be a mismatch of tapers. This should be checked using marking blue and if a good fit is not indicated the new component should be sent back for replacement. (This applies to a new chuck, too, though an error here is unlikely.) A much more likely cause is *dirt*. The two surfaces, inside and out, must be scrupulously clean if the chuck is to be secure – and for it to run truly.

Fig. 43 *Drilling vices. A commercial type (the "Nippy") is at the front, with a massive home-made one behind. The Nippy has a cast iron base but the jaws are hard.*

(A little thin oil will do no harm.) The final and most serious fault is, of course a burr on the mandrel nose of the drilling machine (which can be caused by too enthusiastic use of the wedges in Fig. 41) or on the drill arbor due to bad storage.

The Drilling Vice

It is necessary to hold the work as well as the drill, and though often enough it can be hand-held on the machine table (always with a block of wood or soft metal between), this is not the most prudent of procedures. Especially when breaking through the drill can exert a considerable twisting force and with small workpieces will catch hold, with consequent and sometimes serious damage to the hand. A vice is an essential. However, it need not be a "machine vice" of the type used when milling. Soft, renewable jaws are preferable to hard ones, and the main attributes should be a secure grip, fairly wide opening, and some means of bolting down to the table. It is an advantage if the jaws have vee-grooves, both across and vertical, for holding round stock – it can be a nuisance to have to fit up vee blocks. Most important, the base *should be soft*. With the best will in the world, and even if soft packing is set below the workpiece, sooner or later a drill will be run right through, and if the vice body is hard the point will be destroyed. Fig. 43 shows one good commercial type – the "NIPPY" – and a very rough home-made one which has, nevertheless, served well for 20-odd years. It lacks any holding-down facilities but is so heavy that this is seldom needed! There are many drilling vices on the market, and the only ones to avoid are those made of die-cast light alloy; it is bad enough to have the workpiece spinning round, but if the drill picks up the vice as well, damage to operator, drill, and even the machine is very probable! "Mass" is an advantage in all vices.

SECTION 5

SCREW THREADS

Before dealing with taps and dies it may be as well to have a few words about the threads they produce, as there is quite a number of different thread forms available. A screw is, in effect, a "continuous wedge" wound round a rod, and this feature has been known for many centuries. However, it was not used for fasteners (nuts and bolts etc.) until relatively recently, for with the elementary machinery of a few hundred years ago the simple taper cotter was easier to make and use — indeed, all the blast tuyere pipes on modern blast furnaces are attached this way, not by nuts and bolts; much quicker when a tuyere water-jacket has to be changed. Nowadays the screw thread can be used for (a) Measuring devices; (b) Traversing, as in the lathe leadscrew; (c) Attachments, where components screw into each other (e.g. boiler fittings) and (d) as Fasteners — studs, nuts and bolts. The first two are always screwcut in order to obtain the necessary degree of precision — or possibly thread ground or milled. Type (c) are, preferably, always *started* by screwcutting on the male thread even if finished with a die, but fasteners are, almost universally, made with machine dies in "Production" work,

or by hand for amateurs. Both fastener and attachment threads are, almost universally, of vee-form. The others may be square, trapezoidal, or vee-form.

Fig. 44 shows some of the basic considerations relating to the vee-form thread. Look at (a) and (b) first. Both are the same diameter and the same pitch, but have different thread angles. It is immediately clear that (a) has a much smaller "core" diameter on which to carry the load, so that one might assume that (b) is preferable. However, though (b) provides a greater core area the flatter slope of the vee means (i) that the wedging action will increase the effect of friction compared with (a); this in turn means that for the same spanner effort the fixture will be "less tight". Further, (ii) there will be greater risk of bursting the nut. Not something we think about much these days, but it was not uncommon in the days of wrought iron, and even today when screwing into soft brass so flat a thread could cause trouble. The "best" thread form is, therefore, a compromise between large core area and reasonable thread wedging action. Now look at (c) and compare this with (a). They are the

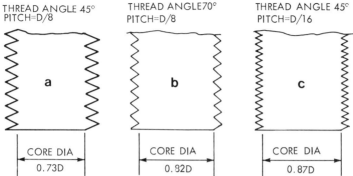

THREAD ANGLE 45°
PITCH=D/8

a

CORE DIA
0.73D

THREAD ANGLE70°
PITCH=D/8

b

CORE DIA
0.82D

THREAD ANGLE 45°
PITCH=D/16

c

CORE DIA
0.87D

Fig. 44 *Comparison of various thread geometries.*

same diameter and have the same thread form, but (c) has a much finer pitch (i.e. more threads/inch) and (c) clearly provides a greater core area. On the other hand, the torque magnification provided by the screw is much greater – in effect, we have a wedge of much finer slope and the tightening stress in the bolt will be greater. In addition, the thread requires much greater accuracy in both form and dimension – a small error will be much more serious. There is also more risk of "cross-threading" in assembly, too. So, again a compromise must be made between the relative attributes to get the best all-round performance. The various standard thread forms we use all effect this compromise in different ways, and, of course, we choose one or the other in making our personal compromise between the types.

Standard Thread Forms

The first engineer to standardise thread-forms was **John Jacob Holtzapffel**, who, as early as 1798 was working on this problem for use on the lathes he made – lathes which, at that time, were well ahead of anything else available. He adopted a "deep" thread of 50° angle,

having sharp crests and roots for "attachments" – mandrel nose, chuck accessories and the like – and a "shallow" thread of about 60° angle, still with sharp crests and roots, for his screws and bolts. He initiated the screwcutting taps and dies over a period and was, so far as I can ascertain, the first to set up a "master system". His methods need not concern us, but having produced an accurate tap and die for any particular size this was used ONLY to make a few "master" taps and dies. These were used in turn to make the "workshop" taps and dies used in production, so that the masters had little wear. However, when wear was found, then a new master could be made from the "originator" tap or die – these might be used only once every ten years or so. It is true that his diameter/pitch combinations may strike us as "queer" – ¾in. × 9.45tpi and 0·36in. dia. × 19·89tpi are examples – but it must be realised that (a) he was concerned only with standardisation within his own works and, (b) this was in the days when the inch was divided into ¹⁄₁₂th, ¹⁄₁₄₄th etc., going down by ¹⁄₁₂th each time from the foot. (And, of course, there were no less than 19 "standard lengths" to the foot in Europe at the time!) These thread types and standards were in use right up until the last lathe was sold in 1925, though quite early on – about 1810 – he changed to a uniform 10 threads/inch for all feedscrews. (A few tiny ones were 20 tpi.)

Joseph Whitworth's approach was quite different. He was aiming at a "Universal" standard, so that a bolt made in Birmingham would fit a nut made in Bristol. At that time (c. 1840) the situation was chaotic. Not only did different makers use different threadforms, pitches etc., but even within the same firm it was not uncommon to find that nuts would fit only the bolts they were made with, as anyone who has dismantled an engine from that period will have found out. He wrote to all the major builders of engineering machinery and collected details of their threads. He averaged the pitches so found at ¼, ½, 1 and 1½ in. diameter, working out at 18, 12, 10, and 6 tpi when rounded off, and used these as a sort of "scale" from which to suggest pitches suitable for the other sizes – and, I suspect, keeping in mind the pitch of the leadscrews of his own screw-cutting lathes! He found that the mean of the thread angles was about 55°, and he took this as his suggested standard. This was quite a steep angle, and to compensate for the loss of core area which would result he proposed the

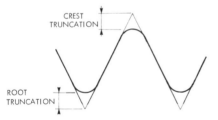

Fig. 45 *A thread form truncated by radii at root and crest.*

truncation of the thread at the root on the bolt; in addition, to reduce risk of damage to the threads at the sharp point (a weakness of the Holtzapffel standard) this was to be rounded off. See Fig. 45. This truncation has, as we shall see a little later, an important effect both on the tapping process and on the behaviour of the threaded pair in service.

In addition to proposing a standard for the threads Whitworth suggested a standard form for the *nuts* so that (a) the number of threads in engagement was such that the various methods of failure were more or less in equilibrium (remember, he was working with cast and wrought iron, not steel), and (b) so that spanners made by one firm would fit all other nuts. The proportions were that the nut height should be equal to the bolt diameter, the across-corner dimension should be about twice the bolt diameter, and hence the across-flats dimension was around $\sqrt{3}\times$ the bolt diameter = 1·732D. These various proposals were presented in a paper before the Institution of Civil Engineers in 1841 and, after some decades, became almost universally adopted, in this country at least.

However, even in 1841 there were some who found that the diameter/pitch combinations suggested by Whitworth were too coarse in the smaller sizes – ½ in. and below – especially those concerned with instrument work, where brass and steel (high carbon steel, that is) were the normal materials. Most of the threads proposed then are now forgotten, but almost all had much steeper angles – 45°, for example – with greater or less degrees of rounding. Whereas ⅟₁₆ in. Whit. was 60 tpi, a common figure for instrument work was 100 tpi. (Whitworth engineering threads

Fig. 46 *Terms and definitions relating to Vee-shaped thread-forms.*

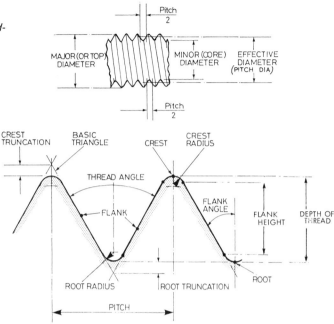

ran down to this size, and rose by intervals of $\frac{1}{64}$ in. up to $\frac{1}{4}$ in.) Nowadays the fine instruments are catered for by the BRITISH ASSOCIATION (BA) thread and watchmakers either by BA or the PROGRESS system.

Current Thread Forms

Before dealing with specific forms it is as well to be familiar with the terms used. Fig. 46 shows the definitions and is more or less self-explanatory. Most of these apply equally to the "nut" – female thread – and the bolt or male thread, but you will not be surprised to learn that the "crest" of the bolt mates with the "root" of the nut and vice-versa! This drawing shows a thread with radiused crest and root, but some, as we shall see, have flats instead.

Whitworth Form

This is shown in Fig. 47, and applies to British Standard Whitworth (BSW), British Standard Fine (BSF), British Standard Pipe (BSP), the British Standard Conduit Thread, and to the "Model Engineer" standard threads as well. The thread angle is 55°, and this gives a height to the basic triangle of $0.96\times$ pitch. The depth of the thread itself is $0.64P$, quite a lot less than the triangle height owing to the truncation, giving a core diameter equal to $D-1.28P$ (D is the nominal bolt diameter) and this applies to all the threads shown. In passing, this factor shows the virtue of a truncated form, as a sharp vee of 55° would leave a core diameter of only $D-1.92P$.

The diameter/pitch combinations are

47

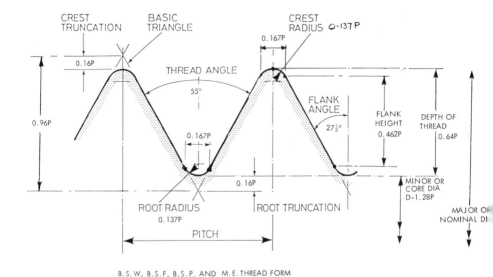

CREST TRUNCATION

BASIC TRIANGLE

CREST RADIUS 0·137 P
0.167P

0.16P

THREAD ANGLE

55°

0.96P

0.167P

FLANK ANGLE

27½°

FLANK HEIGHT
0.462P

DEPTH OF THREAD
0.64P

FLANK HEIGHT

0.16P

ROOT TRUNCATION

MINOR OR CORE DIA
D-1.28P

ROOT RADIUS
0.137P

PITCH

MAJOR OF NOMINAL DI

B. S. W, B.S.F, B.S.P, AND M. E. THREAD FORM

Fig. 47 *The Whitworth thread profile.*

given in the tables at the end of the book; the "Fine" thread, BSF, has now almost entirely replaced the original standard Whitworth except when a stud is screwed into cast iron or a light alloy, when the stud will be BSW at that end and BSF at the nut end. BSF nuts are, in general, one size down from BSW – e.g. a ½ in. BSF nut will have an across-flats dimension corresponding to a ⁷⁄₁₆ in. BSW nut.

British Association (B.A.) Form

This, Fig. 48, is a development of an older thread, the THURY, developed on the continent to suit horological and fine instrument work. Here the problem was that steel screws, often hardened, were screwed into brass plates and were, for their size, pulled up very tightly. A shallow angle was needed to prevent distortion of the female thread in the softer materials. The Thury thread

was essentially metric, but the BA standard is, in fact, defined in inches. The thread angle is 47½°, and is very heavily rounded at both crest and root – the width of the rounded part is 0·236P compared with but 0·167P on the Whitworth. Hence both the depth of thread and the flank height (which carries the load) are less for a given pitch than in the case of Whitworth form. This thread has the ONLY "rational" diameter/pitch combination. The pitch of each size is 0·9× the pitch of the larger one; that for No. 0 BA, the largest, is 1mm; No. 1 is 0·9mm; No. 2 is $0·9 \times 0·9 = 0·81$ and so on – all being "rounded" and expressed normally in inch units. The Diameter of a BA screw is derived from the pitch, $D = 6P^{1.2}$. However, there is no need to work it out, as the figures are given in Table VII on page 91! The core diameter is D−1·2P. This thread is now "obsolescent" and will, in due course,

48

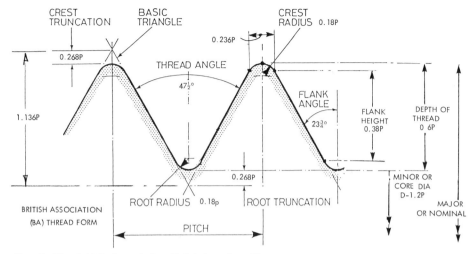

Fig. 48 *The British Association (B.A.) thread profile.*

be replaced by the ISO (International Standards Organisation) series, as will the Whitworth and similar threads. However, the BA series is very useful, and will be with us for a long time yet.

British Standard Cycle Thread

This thread form is now obsolete, as metric forms are almost universal, but as there must be tens of millions of cycles to the old standard I give the

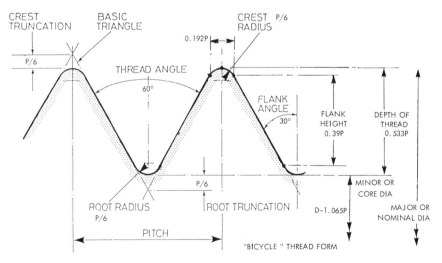

Fig. 49 *The British Cycle Thread. Most cycle threads are now metric.*

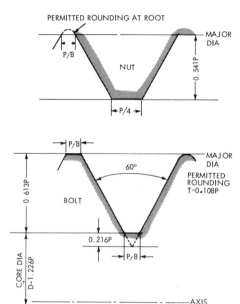

PERMITTED ROUNDING AT ROOT

MAJOR DIA

P/8

NUT

0.541P

←P/4→

→P/8←

60°

MAJOR DIA

PERMITTED ROUNDING T=0.108P

0 613P

BOLT

0.216P

→P/8←

CORE DIA

D-1.226P

AXIS

Fig. 50 *The Unified thread form.*

form in Fig. 49. It has a 60° angle, and is truncated fairly heavily – used on thin tubes it is essential to reduce the core as little as possible – the core diameter is D–1·065P. A useful Model Engineer Pipe thread if taps and dies can be found of suitable sizes!

Unified Thread

This thread was introduced about 40 years ago in an attempt to "marry" American and SAE (Society of Automotive Engineers) thread standards to the Canadian and British usage. It differs little from the original "American National" standard, and in most cases is interchangeable. Fig. 50 shows the form, and it will be seen that the "nut" has a considerable flat on the crest so that it can never "bottom" in the root of the

bolt, and that "permitted rounding" ensures no interference at the root of the nut. It is not dissimilar to the ISO standard shown later, but is, of course, specified in "Inch" units. The core diameter is given by D–1·226P but some users round the root with consequent reduction of core diameter. This form is now the American standard, but is obsolescent in Great Britain, where the ISO form is now standard. There is a number of different series of diameter/ pitch combinations: Coarse (UNC), Fine (UNF), Extra Fine (UNEF), and several constant pitch series which are denoted by the pitch/UN – thus 32UN or 26UN etc. Details of UNC and UNF threads are given in Table XI on page 95.

The ISO Thread

This form was devised by an international committee of the various National Standards Institutions, including B.S.I. All of the thread forms so far considered have disadvantages – either in the difficulty of maintaining tolerances, or accuracy of form, or service problems, and in some cases are prone to manufacturing difficulties in high volume production. The Unified system met some of these, though not all, and in any case was based on inch units. Even in Great Britain it was not too popular. The ISO thread has a 60° angle – Fig. 51 – which can be *generated*, so that accurate form can be achieved without reference to gauges or measuring equipment. The crest of the male thread (bolt) is normally flat, but if rounded the rounding must be *inside* the flat form; and the root of the bolt has a fairly generous radius, giving better fatigue resistance and avoiding stress raisers. The crest of the female thread (nut) has a wide flat, which can never "bottom" on the root

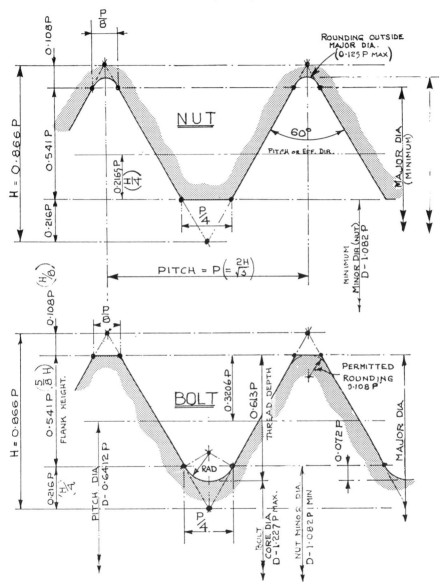

Fig. 51 *The I.S.O. (International Standards Organisation) thread form. It is metric in dimensions.*

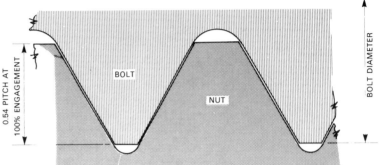

Fig. 51a *Showing the fit of a mating pair to "perfect" I.S.O. profile. The flanks only are in contact.*

of the bolt; the tolerance system (in "production" runs) is such that tool wear on either mating thread cannot cause trouble here. The root of the nut is generously rounded *outside* the major diameter of the bolt so that interference is not possible here either. This has an important consequence in two respects. First, the O.D. of the tap will *be greater than the nominal diameter of the thread*, a point to remember when checking taps with a micrometer! Second, as we shall see later, it affects the choice of tapping drill. Perhaps the greatest virtue of the ISO thread, however, is the fact that it is possible to make a very close-fitting threaded pair *with absolutely NO risk of interference at the crests and roots*. See Fig. 51a. This is very important, and nowhere more so than in model engineering, where we are working with very small screws. The load is taken entirely on the flanks. As can be seen in Fig. 51, the minor diameter of the nut is given by D−1·082P, but the core diameter of the bolt is D−1·226P.

Diameter/Pitch combinations

With the exception of the B.A. thread and the British Standard Pipe Thread there are various combinations of dia-

meter and pitch available with each system. The aim is to maintain a more or less constant angle of thread in any given set – Coarse or Fine. The use of a finer pitch for a given diameter does, as already noted, provide a larger core area. In addition, because the fine pitch involves a smaller circumferential angle to the thread, a given spanner torque will apply a greater tightening load in the bolt. There is no difference in the strength of the actual thread, no matter how fine it may be, but clearly the smaller the thread section the more important it becomes to avoid dimensional error and the more serious is any damage to the crests. Anyone who has dismantled an old telescope, where the threads are very fine, will realise the difficulty that can be experienced just due to thermal expansion! There is a final point; "classical" engineering practice is to have between five and eight threads engaged in any threaded pair – bolt and nut, or whatever. When using attachment threads into relatively thin members the use of a finer pitch was obligatory. So, the choice of fine, medium or coarse pitches is a question of purpose and design rather than manufacture or strength.

Constant Pitch Series

Of these the industrial standard most useful to model engineers is the BRASS THREAD, of 26 tpi – Table VI. This was used mainly on thin-walled brass tubing for the old gas-brackets, and was more fitted to the thin piping than the corresponding "Gas" thread, intended for thicker wrought iron and, later, steel tubes. The constant pitch enables different sized pipes to be coupled together by screwing the adaptor coupling well onto one pipe and then bringing it forward onto the other. Unlike the B.S.P. (British Standard Pipe) thread, the pitch is the same for all sizes, whereas B.S.P. has varying pitches in the smaller sizes (Table XIII). It is superseded by the I.S.O. 1 mm pitch series.

The MODEL ENGINEER threads, of Whitworth form, were introduced after many decades of debate and confusion about 40 years ago; in this case the need was both for a finer pitch than then currently available, and for a constant pitch which could be used in making small fittings. Originally 40 tpi was used for sizes up to ¼ in. and 32 tpi above this. However, the 40 tpi range has been slowly extended upwards, and can be had (at a price) up to ½ in. diameter. There is great risk both of cross-threading and of bad fits above about ⅜ × 40, and it is only in rare cases that even this should be necessary. More recently one enterprising manufacturer has re-introduced the 60-tpi series in diameters from ¹⁄₁₆ in. up to ³⁄₁₆ in. These are very useful for adjusting purposes and if one end of a rod is screwed 60 tpi and the other 40 tpi a handy "differential" adjustment is possible. These threads are given in Table VI.

The MODEL ENGINEER'S METRIC ATTACHMENT THREADS, Table X, were drawn up in 1982 by a committee of eminent model engineers appointed by the publishers of *Model Engineer*. The object was to devise a series of threads which would both accord with the I.S.O. standard and at the same time meet the special needs of model makers of all branches. It had taken nearly 50 years for the Whitworth form M.E. threads to be agreed and adopted, and it was felt that although it might be some time before these (M.E.) screwing tools became very expensive it would be as well to have a metric standard which manufacturers could work towards. The series was found acceptable to the British Standards Institution and is, in fact, published by BSI as P.D.6507/1982. There are three standard pitches, 0·5 mm, 0·75 mm and 1·0 mm, with some overlap of diameters in each set. The 1 mm pitch series is particularly useful for large scale workers, as it forms part of the I.S.O. 1 mm series, running right up to 20 mm dia.

It may be some years before this standard comes into use, but it will be there when needed and it is to be hoped that designers will use it, and so avoid the disastrous confusion which existed in former times. It is worth mentioning that the same working party drew up recommendations for the hexagon sizes for use with I.S.O. threads in models – those used for industrial purposes are far too clumsy – and P.D.6507 gives details of these also, as does the "Model Engineer's Handbook"**.

**The author was the Secretary to the Working Party which drew up these standards.

SECTION 6

TAPS AND DIES – GENERAL

Before dealing with these in detail a few general points are worth discussing. The most common question asked is whether there is any advantage in using High Speed Steel against the cheaper Carbon Steel for both taps and dies. The first point to note is that industrial needs are very different from those of the amateur or even the jobbing workshop. In industry hand tapping is very rare and the use of a die by hand almost non-existent. Tapping is done with tapping machines which have a delicate torque control to avoid over-twisting, and which automatically reverse the rotation to withdraw the tap on reaching the preset depth. These machines work fast, and are working all day long, so that while it may seem inconceivable

that a tap can get hot it may get at least "well aired" in such service! Similar considerations apply to automatic screwing machines making male threads, though these use, as a rule, special die-chasers rather than the button dies used in the hand die-stock. For hand work the hot-working feature of high speed steel is of no relevance. Carbon steel is, in fact, *markedly harder* than H.S.S. and this is an advantage in hand work.

More important than the material is the way these taps and dies are made. In the case of *carbon steel* the tap or die is made either by screw-cutting or from a master tap or die, fluted, and *THEN* hardened, the edges finally being sharpened by grinding the faces in the flutes.

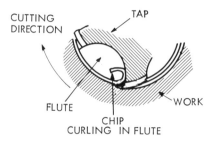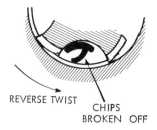

Fig. 52 *Showing how the chips in the tap or die flute may be freed by reverse rotation.*

54

Even though very special furnaces are used and close control is kept over both hardening and tempering there is some risk of distortion. The tolerance allowed on the straightness has to take into account the manufacturing methods, as does that on the thread form. With some cheap taps especially quite a marked bend in the shank may be found. *High-speed steel* taps and dies on the other hand, are made from hardened blanks. The hard barstock is first ground to diameter, the thread is then formed by thread grinding and then finally sharpened in the flutes. Thus the whole of the thread must be co-axial with the shank, the thread form will be to a finer tolerance (thread-grinding wheels are continuously monitored and automatically reformed when needed) and – of importance to those who have Quorn cutter grinders – the male or female centre at the business end can be relied upon when resharpening becomes necessary.

For jobbing work, therefore, a H.S.S. tap or die will cut more accurate threads, but will not be as hard as one of carbon steel and (under hand-use conditions) is likely to blunt sooner. H.S.S. is marginally better in its resistance to chipping of the cutting edge. I have found no advantage either way over the problem of tap breakage but if a tap *does* break then there is no comparison – carbon steel is far easier to remove as it can, in the last resort, be softened with quite moderate heat! H.S.S. is, of course, more expensive.

Cutting Action

The cutting action of the teeth of both taps and dies is the same as that of a single-point screw-cutting tool, but the teeth are so arranged that each takes out a little more of the total depth. Once fully engaged (in the case of a tap, when it is one diameter deep) the load in the shank and in the workpiece is much greater than would be the case if the thread had been screwcut full depth with a single point tool, due to friction. In H.S.S. ground thread taps and dies of any size – above, say, 8mm (5/16in.) the teeth have circumferential relief to reduce this, but the torque is still considerable. We shall cover this further when we come to the selection of tapping drills.

A more important point is that with ductile materials (worst of all with bright-drawn steel) there is far less room for the chips. Though theoretically the tap or die is designed to provide ample space in the flutes or clearance holes, curly chips do tend to jam up – see Fig. 52. The rule is to make only half a turn forwards followed by half a turn backwards to break off the chips; if this is not done with all ductile materials sooner or later you will find that the tap or die cannot be moved either way – though the problem is not so great with dies as, with luck, the chips may curl right out of the clearance holes. With free-cutting materials which do not produce this type of chip the problem is reduced and with "screwing brass" the thread can be cut continuously. When making male threads in the lathe free-cutting mild steel and most non-ferrous metals can be threaded with a die from the tailstock under power, but unless bright-drawn mild steel is first normalised the threads are likely to tear.

One final point which applies to split dies (see page 72). All dies are designed to cut the thread to full depth in one pass. The split feature is intended to allow the dies – especially those of carbon steel – to be adjusted to size.

Whilst it is permissible practice to take a *fine* second cut to adjust the size of the thread to gauge, the habit of "roughing down" with the die open, followed by a "finishing cut" should be avoided. This causes unnecessary wear. If it is found that the die *has* to be run through twice because the effort with a single pass is too great, then it needs sharpening, and this should be attended to. Frankly, I prefer to use solid dies, but these are usually obtainable only in H.S.S. with ground threads. Kept sharp, they produce a thread of the correct size every time – especially important with threads below 3mm (5BA).

SECTION 7

HAND TAPS

Fig. 53 is a generalised illustration of a typical hand tap. The body is threaded and has from two to five flutes cut longitudinally, depending on size, though 2-flute taps are usually for use on tapping machines only. Depending on the shape of the fluting cutter the rake of the cutting face may be straight or curved, as seen in the top RH figures. For production work these rakes are carefully specified according to the duty, but this need not concern the jobbing worker. As already remarked, taps may be cut with thread relief, but this is seldom found on carbon steel cut thread taps. Such relief eases the cutting loads – reduces friction – but on the other hand a tap without relief is better guided by the threads already cut. Unless a lot of work is to be done in difficult materials – work-hardening stainless steel, for example (or uranium!) – it is not worth the extra cost to ask for this feature. Small taps will have male centres at the point and may have a similar point at the

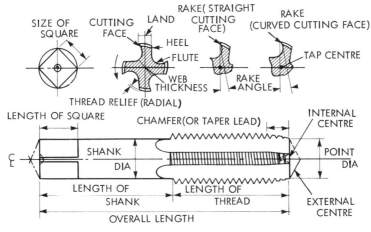

Fig. 53 *Tap nomenclature.*

Courtesy B.S.I.

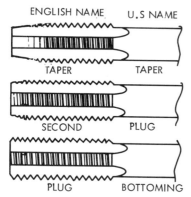

ENGLISH NAME   U.S NAME

TAPER   TAPER

SECOND   PLUG

PLUG   BOTTOMING

Fig. 54 *The shape of taper, second and plug or bottoming taps.*

squared end, though small H.S.S. taps are held, during manufacture, in collets on the shank. Larger taps may have either male or female centres at the threaded end. Male centres often get in the way when tapping shallow blind holes; before grinding it off remember that you will then lose means of holding the tap at both ends when resharpening in the Quorn or similar machines.

The entry end is chamfered with a taper lead, and this "lead" is intended to start the tap into the hole. As seen in Fig. 54, there are three lengths of taper. The longest has an included angle of about 8° and the end is smaller in diameter than the usual tapping drill;

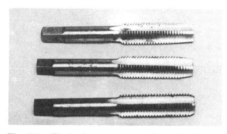

Fig. 54a *Typical taper, seconds and plug or bottoming taps.*

there may be 10 to 15 threads chamfered off. The "Seconds" tap has a shorter taper of about 16° included angle, whilst the "plug" or "bottoming" tap is shorter still, about 45°-50° included, which means that in effect only a single thread is affected. (Note that in USA the term "Plug tap" is applied to the British "Seconds" type; it is better to use the term "Bottoming"). Occasionally one will find a bottoming tap with no lead at all, presumably with the intention of carrying the thread right to the very bottom of a blind hole. There is little point in trying to do this and the cutting action of the leading thread of such a tap is so poor that the leading (cutting) edge soon blunts or, worse, chips off.

The taper tap is intended for making the first cut in through holes – it is useless on a blind hole. The threads even on the parallel part are undersize, and it must therefore be followed by the seconds tap. This will give a full thread if taken beyond the taper, and there is no need to follow it with a bottoming tap. For blind holes, however, the *seconds* tap is used first, followed, *only if necessary*, by the bottoming tap if the seconds does not take the thread deep enough. **For general purpose work the seconds tap will cover almost all needs.** The main point of the taper is that it reduces wear on the seconds and, in the case of very small taps, reduces the risk of breakage as there is less load on the seconds tap when it takes its turn.

Occasionally you may meet with taps which are made with *flats* rather than flutes in very small sizes (below 12BA or 1·5mm). These have either a square or triangular section and, clearly, operate with considerable negative rake on the cutting edges. These are suitable only for brass, and even there must be used with great care. It is almost impossible

Fig. 55 *A "Nut Tap" showing the threaded nuts feeding along the shank.*

to use them for blind holes – the use of the taper tap is almost imperative.

The Nut Tap, Fig. 55, is intended for short production runs making nuts, and as can be seen in the photograph the shank is reduced to less than the core diameter of the thread, so that when tapped the nuts can run over this part. That shown has a square on the shank, but this has been ground on to convert it to a hand tap. They are strictly *machine taps*, having a plain shank with a single flat for holding in special collets. The lead is longer than a seconds hand tap and usually the flute is of special form to give free cutting at high speed. Many taps sold as "surplus" are of this type and very useful they are, too. However, they may need resharpening if they have come from a production workshop. You MAY find that nuts made with such a tap can be fitted only with difficulty onto a bolt; you have picked up a tap to Class 1 tolerances, too close for normal work!

Spiral Point Taps have the flutes formed into a helix (a "spiral" tap would get you nowhere very fast!) with the intention of feeding the chips forwards out of the flutes. This certainly makes for easier tapping, but there is a price to pay. The chips will be fed to the bottom of the hole and will not remain in the flutes. This means that you will not be able to reach the bottom of a blind hole without first blowing out all the chips.

They are intended only for through holes.

There are many special taps of one sort or another, and these turn up both at club jumble sales and on the surplus market. They are best left alone, or kept as curiosities. Almost all of them will be machine taps, intended for use with tapping machines; they may be to special or unusual tolerances, or have obscure rake angles intended for awkward materials and which will jam up if used on steel or brass. The "common or garden" hand tap, whether of H.S.S. or carbon steel, will meet almost all needs. The only "specials" I have needed over half a century of making screwed holes have been for odd pitches or to reach awkward places, and I will deal with this latter problem in a moment!

Tap wrenches

This is an unfortunate term, as the last thing you should do to a tap is to "wrench" it! The term arose, of course, in the days when a half-inch thread was regarded as "small". The important thing is to have means of turning the tap smoothly, with even torque so that there is no bending action, and of a size which "matches" the size of the thread. In the old days this would be a bar of flat steel with square holes in it to fit various size tap shank. A barbarous affair; I have seen them with holes suitable for 10mm (⅜in.) taps down to

59

Fig. 56a *The so-called "American" type tap wrench; this one will accept ½ in. squares.*

3mm (5BA) all on the one bar – asking for trouble.

Fig. 56 shows various acceptable types. At 'a' is a large one, accepting up to 12mm (½ in.) squares, say 22mm or ⅞ in. taps. It is about 40mm (15 in.) long progressively smaller down to about 5mm or ³⁄₁₆ in. The lowest is a chuck type, for taps from 4mm or 5BA down to 2mm or 8BA. One point to note: in the upper three the tap square should be set in the hole so that the locking

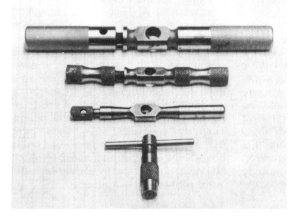

Fig. 56b *Three bar-type wrenches; the largest will accept a 12mm or ½-inch Whit. tap. Below, a small chuck-type to take up to 4BA.*

and strong in proportion. This type of wrench is sometimes known as the "American", but the one shown is British to the last knurl! In Fig. 56b I show four types, the largest at the top for taps up to 12mm (½ in.) dia. and the two below screw bears on a *corner*, not on the flat. The largest wrench would be improved if the circular hole were made more like that of the lower type. Finally, at Fig. 56c I show the "wrench" I use on very small taps – 12BA or 1·5mm. It is not my

Fig. 56c *A pin-vice, used as a tap-wrench for smaller taps. That shown is 12BA.*

60

Fig. 57 *Showing how a pin-vice may be used to reach an inaccessible hole. The tap is 4BA.*

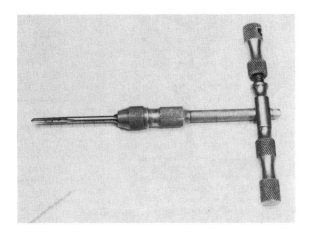

smallest pin-vice though, and I have one with jaws which will hold down to 18BA. In most cases enough grip is had from the fingers, but it is important NOT to turn it using the knurls at the nose, or this will unscrew and release the tap. Finally, in Fig. 57 I show how the shorter taps of small size may be caused to reach down into awkward places. This is a larger pin-vice than that in Fig. 56 and is holding a 4BA tap (3·6mm) with one of the medium size (Fig. 56b) wrenches on the end of the vice.

Keeping taps square to the work

This is a perennial matter for discussion! Like many others, I was taught to use a small trysquare once the taper had started to cut and to ease it one way or the other to rectify any "slantendicularity". Once the taper tap runs square, of course, the others will follow. Many devices, simple and complex, have been described to ensure squareness. All need making and many need setting up. For those who do a great deal of tapping into blocks the "George Thomas" Staking and Tapping Tool (or "Pillar Tool") is the answer (Fig. 58) but

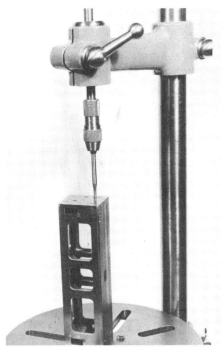

Fig. 58 *The George Thomas Staking Tool, seen being used to guide an 8BA tap. (Courtesy of G. H. Thomas Esq.)*

61

Fig. 59 *The author's method of guiding taps. A similar method can be used in the lathe (see text).*

it has to be admitted that it is a major piece of equipment and hardly worth while unless there are other uses for it – it is, after all, a STAKING tool. But it will handle taps from ⅜in.×40tpi down to 16BA, a range not to be despised! However, my own arrangement is effective for 85% of my work, shown in Fig. 59. The workpiece is held square in the vice on the drilling-machine table and the tap guided by the drill-chuck which is finger-tightened until the tap can just be turned. The drilling machine spindle is, of course, locked. The wrench is fitted to the shank and most of my taps now have a *small* flat ground in a convenient place. The same arrangement is used with the tailstock drill-chuck in the lathe. It is effective and has the merit of costing nothing – you must have a drill chuck before you can tap a hole, anyway! The one limitation is that it cannot be used with taps which are not reasonably straight, but this applies to any other method. (The operation of "Murphy's Law" has ensured that my best-cutting ¼×32 tap has a badly distorted shank!)

SECTION 8

TAPPING DRILLS

At first sight it would seem that all we have to do is to drill a hole the same size as the core diameter of the thread and then use the tap. This is far from being the case. First, the cutting forces are so great that large taps would be almost immovable and small taps too weak to stand the torque. But there is another factor – one which you can try for yourself. Choose any size tap about 3 or 4 mm (say ³⁄₁₆ × 32) and use a drill which is about three-quarters of the O.D. of the screw. Say 3.6 mm for ³⁄₁₆ × 32. Drill carefully – make sure that the drill is properly formed; use a new one if you can. Tap the hole equally carefully. Now try the drill to see if it will still enter the hole. If you have drilled in steel or any ductile material you will find it just won't go in, and even with brass it may be a stiff fit. The reason is that even with a sharp tap the cutting forces tend to *extrude* the metal into the vacant space at the root of the tap threads. If the hole is core diameter size there will be nowhere for this metal to go; it will be compressed and the friction on the tap will, even at this size, be sufficient to break it. This means that we **MUST** use a tapping drill which is larger than the

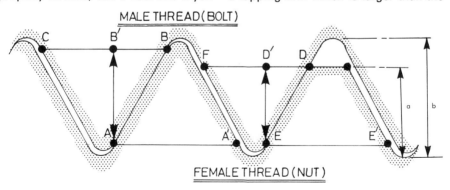

Fig. 60 *A threaded pair shown in full (left) engagement and (right) in partial engagement. Note that only the flanks of the threads are in contact in both cases.*

core diameter, and quite a bit larger at that. The result will be what is known as "Partial Engagement" of the threads. Look at Fig. 60. On the left is shown a mating pair — nut and bolt — with full engagement of the threads. Apart from the unavoidable clearances the male thread (bolt) fills the female thread completely. On the right hand thread, however, the male thread is complete but the female thread in the nut does NOT fill the male thread completely. The ratio a/b, expressed as a percentage, is called the "**% thread engagement**". On the left of Fig. 60 this ratio is 100%, on the right about 75%. This latter is about the normal for power-driven machine taps in industry, and is the basis for most of the tables published by tap and die makers, B.S.I., etc. But remember what I said earlier; these machines have sensitive torque limiting devices, and they slip well before there

is any risk of the tap breaking. Further, the taps they use are all really sharp — they are taken out of service as soon as their performance begins to fall off. How long is it since you sharpened any of *your* taps? I think you will find that if they were put into a tapping machine they might well start slipping straight away!

Now look at Fig. 61. This shows the relative torque required to drive a tap at various % thread engagements. (It is drawn for Unified Thread taps, by the way; the effect will be more pronounced for BA or Whitworth form threads.) You can see that at 75% engagement the risk of tap breakage is 2·3 times that at 60%; at 80% it is three times as great, but if we went down to 50% the risk would be halved compared with 60%. Or, put the other way round, if we call the risk with industrial thread engagement as 'X', then at 65% the risk is

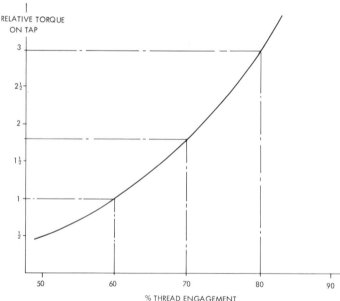

RELATIVE TORQUE
ON TAP

% THREAD ENGAGEMENT

Fig. 61 Graph showing how the tap torque (and hence the risk of tap breakage) increases as the depth of thread engagement increases.

0·59X, at 60% it is 0·43X, at 55% about 0·3X and if we went down to 50% thread engagement it would be only 0·22X. For the jobbing worker and the amateur this is clearly a serious consideration, for whereas an industrial firm will buy taps in dozens (or tens, these days) and always have plenty in stock, you, like me, may have only one of each size – AND no tool-dealer within 30 miles from whom to get a new one! So, does it matter if we use a low engagement figure?

Strength of Partially Engaged Threads

Look again at Fig. 60. The first thing to notice is that we gain no benefit from 100% thread engagement anyway. The threads not only do not, but MUST NOT, touch on the radiused crests and roots; if they do they will jam. The load is, in fact, carried ONLY on the flanks – from A to B on the left hand side. The TRUE depth of engagement is not a/b but DE/BA. This actually works out at 82% on the diagram instead of the nominal figure of 75%. Now consider what happens if the threaded pair fails. First, the bolt may snap off across the core diameter. This is, in fact, the basis on which the bolt will have been designed. (I use the term "bolt" to indicate ANY male thread, whether it be a fitting, adjusting rod or actual nut and bolt.) Or the thread may "strip". This means that it will *shear* across the line BC on full engagement, or DF with partial engagement. The third method of failure might be in *crushing* of the flanks – along AB or DE.

I won't bother you with the mathematics or the "materials science" of the thing; suffice to say that at 100% engagement the bolt thread is *three* times as strong in thread shear as it is across

the core area in tension, and no less than *five* times as strong in resisting crushing of the flanks. (This assumes the full nut height and reasonably well made threads, of course, corresponding to the coarsest commercial fit.) The nut, or female, thread is even stronger in shear, by the way, if it is made from the same material as the male thread. One reason why we can use mild steel nuts on high tensile bolts! However, the point I am trying to make is that all we do when we use a low thread engagement is to *change the ratio of shear to tensile strength*, and we should have to lose two-thirds of the shear area before the strengths become equal.

In Fig. 62 I have analysed one of the most common threads we use – the BA form. I have, in effect, sliced off the *nut* threads at intervals of 5% of the total height and worked out first, the actual flank engagement and then the "strength ratio", assuming that bolt and nut are of the same material and that the nut is one bolt diameter tall. You will see from the table on the right of the diagram that there is NO POINT in using a thread engagement above about 82% – you gain nothing in strength but the torque curve on Fig. 61 goes off the chart. At 60% thread engagement the strength ratio is 2·5 – i.e. the threads are 2½ times as strong as the core diameter of the bolt. *NOTE THAT THIS IGNORES ANY EXTRUSION EFFECT* – the *actual* thread engagement will always be about 5% more than the nominal, more in some materials, especially BDMS. I have shown the results in graphical form in Fig. 63, and have added a second curve which shows the comparison between the *nominal* thread engagement and the engagement of the load-bearing flanks.

What this all means in practice is that

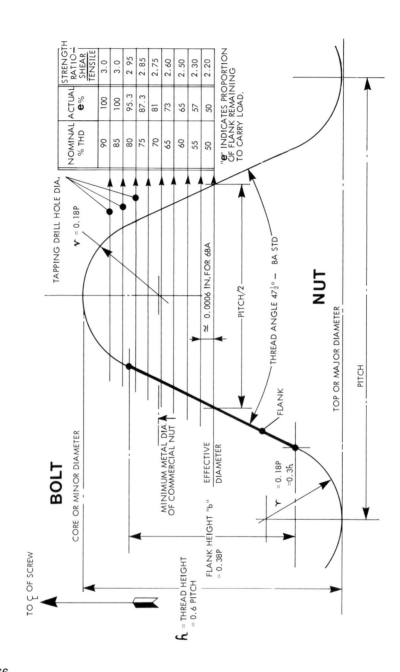

Fig. 62 *Analysis of various degrees of thread engagement for a BA threas. See text.*

Fig. 63 *Graph showing (a) true % flank engagement and (b) relative shear/tensile strength of a bolt based on the conventional figure derived from the drill size, BA thread form.*

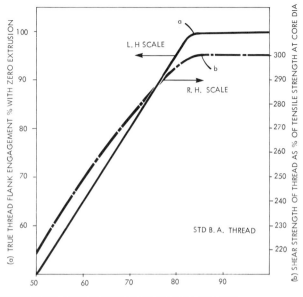

CONVENTIONAL THREAD ENGAGEMENT% BASED ON DRILL SIZE

simply by going down from the conventional 75% (or more) thread engagement to 65% we HALVE the risk of tap breakage with a loss of no more than 7% of the shear strength of the thread. If we go down to 60% the tap breakage risk is reduced still further and the loss of shear strength is still only 12½% – and still 2½ times as great as the tensile strength of the bolt itself. I have for over 10 years used a general figure of 60% for threads below 2·2mm (7BA), 65% for those between 2·2 and 6mm (¼in.) and between 70 and 75% for larger diameters with no ill-effects at all – and my tap breakages have fallen dramatically. Indeed, below 8BA I am now having to *sharpen* taps; previously they tended to get broken before they got blunt!

This calculation has been done for the BA thread; identical considerations apply to those of Whitworth form. In the case of ISO metric threads the situation is even more pronounced. Look again at Fig. 51, page 51. You will see that the *tap has a greater diameter than the male screw thread.* If we calculate the "nominal engagement" of the threads in the ordinary way, based on the bolt thread height, the tap will actually "see" a greater engagement. Thus, a "nominal" engagement of 65% will provide about 74% of flank height, but the tap will find itself with about 76% "cutting engagement". It is even more important with this type of thread form to examine the balance between the risk of tap breakage and the strength of the thread at reduced engagement.

Other considerations

We have so far assumed that the hole drilled is to proper size. But take a look

67

again at Fig. 62. In the case of a 6BA (2·8mm) thread the difference of 5% engagement represents a hole size difference of only 0·0006in. – 15 microns. It is highly unlikely that anyone can drill to that sort of tolerance using old jobber's length drills resharpened by hand. I have for many years used *nothing but stub length drills* for tapping – few tapped holes are of any depth. They are kept in a special box (see Fig. 37, page 35) and *I use them for no other purpose*. I strongly recommend this practice though, of course, within reason; the possible error ceases to be of significance with threads above 5mm or ¼in. diameter.

When tapping *thin material*, threaded right through, the risk of tap breakage is reduced, as even on the seconds tap the full-thread cutting edges will not engage until the leading threads have emerged from the underside. Further, we now have fewer threads in engagement, and although the strength in shear is seldom a consideration in such work it is worth aiming at rather deeper thread engagement. I usually use between 5% and 10% more than I would normally – again, note that Fig. 62 shows that for BA threads there is no point in going above 80% nominal.

Certain **materials** are easier on taps than others, and **cast iron** (the softer grades), **cast aluminium**, and **free-cutting brass** ("screwing brass"), can be worked at least 5% higher engagement than those I suggest on page 65. On the other hand, **stainless steel, chromium alloy steels**, and **manganese bronze** should be worked between 50% and 55% only. I have found that **monel metal** can be a bit tricky, and some grades of **German (nickel) silver** can cause trouble. Both are strong materials and 5% less engagement will do no

harm. In all these difficult materials the sharpness of the tap can be crucial and this is nowhere more so than in the case of work-hardening grades of stainless steel.

Tapping Drill Tables (See pages 89 to 101)

There are two approaches to selecting tapping drills, and I have tried to cover both in these tables. The first is to decide on the thread engagement needed and then look up the size of drill necessary. The second is to have a standard set of drills and use a table to find out what engagement each will give. The second procedure is covered by Table XV (page 99) where I have listed the modern standard metric set running from 0·1mm to 10mm×0·1mm. Thus, if you want to tap 7BA, look down the BA column and you will see that 2·0m gives 82% and 2·1mm 65% thread engagement. Take your pick!

Alternatively, say, for BA threads use Table VII, page 91. This gives the FULL details of the thread and then a choice of drills for different thread engagement. Those shown in bold type are the British Standard "preferred sizes", but ALL the sizes shown are available in jobber's drills. Stub drills in the smaller sizes do have a more limited choice, but some sense of proportion must be maintained! The difference between 1·89mm and 1·90mm for a 60% engaged 8BA thread is hardly worth bothering about! (Four-TENTHS of a thou!) If the "words and music" call for a 70% No. 0BA thread (look at the table) then the use of a 5·1mm or 5·2mm drill will not bring the world to an end. Indeed, if the 5·1mm is old it may well drill 5·2mm anyway. But I have given the size of drill in ALL the tables which is the nearest available to provide the depth of engagement shown.

The sizes between bold vertical lines are those which I use. They are stocked by several Model Engineers' materials suppliers.

There is also, in all the tables except those covering metric threads, a column for "General Purpose" tapping drills. These are all in the "old tradition" – fractional or Morse number and letter drills. These will provide a thread engagement somewhere between 60 and 75%. But if you want to work closer, just look up the *metric* size of drill needed and then use Table III, page 87, to find the nearest number or letter drill. I have not done this for the metric threads because if you ARE going metric with taps and dies you should do the same with your drills. Finally, at the head of each table there is a formula which will enable you to calculate the drill size should you need any depth of engagement for some special reason.

The tables for *ISO METRIC THREADS* are rather different in one respect. We have already seen that the O.D. of the tap is greater than that of the male thread. This means that if a drill is selected to give (say) 70% of the male thread engaged it will provide about 80% of the flank height and the tap will actually "see" about 82% when it is cutting. So, in this case the column headings are based on *flank engagement* and I have shown in square brackets the engagement seen by the tap. You will see that the difference is slight, but as it is there I have thought it wise to give you the data.

One small point about the threads in Table VIII. This includes the ISO Miniature Instrument threads to B.S. 4827 – the 'S' range, as well as the standard ISO range. Note that although S1.0 and S1.2 have the same pitch as their M1.0 and M1.2 equivalents they are NOT interchangeable. M-type nuts will not fit S-type bolts, and though S-type nuts may appear to fit M-type bolts the thread mating will be poor. The two types should be kept separate.

Finally, for those concerned to replace BA screws with their ISO counterparts I have given a rough comparison in Tables VIII (ISO standard) and IX (ISO fine pitch) in the final column. These comparisons are based on the thread O.D., though in most cases the core area is about the same, too.

Jointing threads

Many practitioners have the idea that if they use almost 100% thread engagement they can rely on making a fluid-tight joint in the threads. This is just not true, not with tap-and-die cut threads anyway. The commercial class fits even when using ground thread taps and dies always leave a clearance on the flanks and at the roots of the threads, and even when, as in full-size pipe-fitting, taper threads are used and pulled up really tight, some form of jointing has to be used. It is far more effective to concentrate on cutting a good, clean, thread profile and then to use an appropriate jointing compound. In the special case of boiler stays this is even more the case. Copper is liable to tear if too small a tapping drill is used and, in addition, it has, in the annealed state after brazing, a greater tendency to "extrude" than most metals. If a 60 to 65% drill is used the threads will be well formed for strength – and you will be "caulking" with soft solder or, better, brazing alloy to make the seal anyway.

Clearing drills

I have heard criticism of some published drill tables to the effect that they do not give the "right" size of clearing drill for each size of thread. This is a puzzler! A number 0BA thread is 6mm O.D., so that a drill 6·05mm dia. will give a clearing hole. But so will a 7mm – or even 10. It all depends on what you want the clearance FOR. Looking at published drawings over many years my own view is that designed clearance holes are usually too small for the application. Even in the old days I always used one, and often enough two, sizes of drill larger than that called for on the drawing for such things as cylinder covers, valve-chests and the like. However, the criticism had to be met, but I have tackled it another way! In each table you will find the O.D. of the thread given *in millimetres*. This means that you can select your clearing drill directly if you have metric drills and only need to refer to Table III (page 87) if you are still stuck with number and letter drills. But keep in mind the job that the hole has to do; if there is but a single bolt to fit, then quite a small clearance – 0·05mm – should serve. But if you are going to have to wangle a cover over ten or a dozen studs allow quite a bit more, even if they have been jig-drilled, otherwise you will spend a lot of time trying to get it (the cover) on.

Conclusion

Tapping drill sizes as used in industry are not appropriate to jobbing work, especially in small scale. Even with 50% thread engagement the loss of shear area is unimportant unless the actual *number* of threads engaged is small (less than four or five). Between 60% and 65% is adequate for almost all applications, and these proportions will reduce the risk of tap breakage very considerably. On the other hand, in small scale work the difference of 5% thread engagement is very small, so that it is very important that drills cut to size. Stub drills, especially if ground with four-facet points (see page 34) are far superior to jobbers for tapping work. Finally, if at all possible – and I urge you to consider "doing without" something else to MAKE it possible – keep your tapping drills for that purpose *only*, and try not to use them for anything else.

SECTION 9

SCREWING DIES

The earliest form of commercial tackle for making screws and male threads was the **Screwplate**, Fig. 64. At 'A' is the "Warrington" pattern used by clock-makers, but that at 'B' is the more usual pattern. The "Warrington" has a number of gauge-holes for checking the size of the wire before screwing in the matching die-hole. That at 'B' carries roughing and finishing die-holes. For small screws the blank would be driven through using the screwdriver slot or hexagon head, but for longer threads the tool was rotated using the handle. About the middle of the 19th century the die-stock with interchangeable dies was introduced; Fig. 65 is from the Britannia catalogue of 1890. The principle is easily followed from the drawing; the two half-dies were set in the recess and screwed up tight – the pinching screw is NOT for adjustment. (The apparent ovality is due to the way they were made – the actual threaded part is circular.) By fitting a pair of vee-shaped dies the same tool could be used to hold taps. Even in those days such tools were expensive – a set covering ⅛in. to ⅜in. Whitworth cost £2.70 at a time when a fully equipped 4in. centre-height screwcutting lathe could be had for £25!

The Button Die. This is the universal style these days, except for threading large diameter pipes. There are two types. The SOLID die is shown in Fig. 66. For the smaller sizes, when the die is less than 25mm or 1in. dia. there are either three or four cutting edges, but for larger dies of 25mm (1in.) O.D. upwards there may be five or more. These dies cannot be adjusted for wear, and so long as they are sharp will cut to size. However, for jobbing and amateur

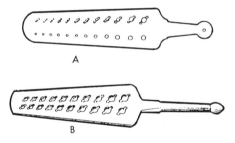

A

B

Fig. 64 *Illustration of 19th century screwplates taken from an early textbook. Upper, the "Warrington pattern, with gauge holes for the screw-wire. Lower, typical "Engineer's Screwplate" for smaller sizes, with roughing and finishing holes to each size.*

BEST STOCKS & DIES,

FOR ENGINEERS' USE.

☞ Whitworth's Standard Thread.

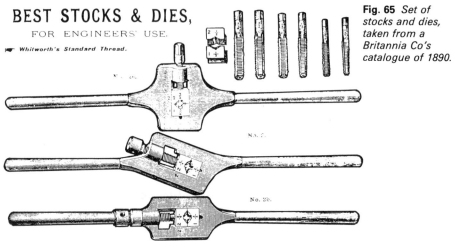

Fig. 65 *Set of stocks and dies, taken from a Britannia Co's catalogue of 1890.*

use the SPLIT DIE is more common – though, as explained earlier, many users are changing to the solid type. A range of split dies is shown in Fig. 67, and it will be seen that the largest – 1½ in. dia. – has only four cutting edges. This is non-standard.

Such split dies clearly need means of adjustment. At one time they were available with an adjusting screw within the die itself, but sheer cost has driven them from the market; any found as "surplus" or at sales are likely to be

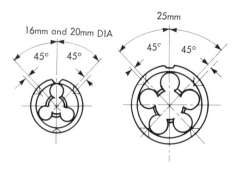

Fig. 66 *The modern solid pattern button die.*

nearly 50 years old or more and should be treated with reserve, as they may have had a great deal of use. However, the small ⅝ in. dia. dies are still made and these can be fitted into collet holders, as shown in Fig. 68. These collets will not fit the present standard die-holders – they are 0·877 in. O.D. – but a ¹³⁄₁₆ in. holder can easily be adapted. The collet is hardened and the pinching screws lie below the circumference. A useful feature, seen in the top left-hand view, is that the front of the collet has a guide to direct the workpiece axially to the threading teeth. In all other types of die it is necessary to have means of adjustment in the die-holder and to adjust each time the die is used. We will come to this point later.

Dienuts

These are intended for the restoration of damaged threads on bolts, and at first sight appear to be no more than a solid die with a hexagon outline to fit a spanner. They are simply screwed onto the bolt and worked along the length of

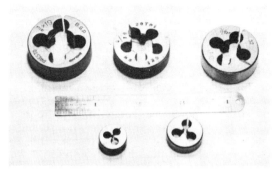
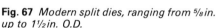

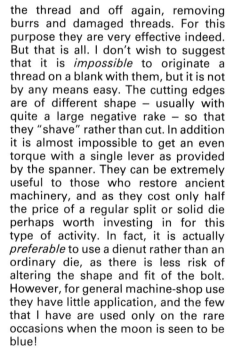

Fig. 67 *Modern split dies, ranging from ⁵⁄₈ in. up to 1¹⁄₂ in. O.D.*

Fig. 68 *The "Pratt & Whitney" type die, normally fitted in a collet holder.*

the thread and off again, removing burrs and damaged threads. For this purpose they are very effective indeed. But that is all. I don't wish to suggest that it is *impossible* to originate a thread on a blank with them, but it is not by any means easy. The cutting edges are of different shape – usually with quite a large negative rake – so that they "shave" rather than cut. In addition it is almost impossible to get an even torque with a single lever as provided by the spanner. They can be extremely useful to those who restore ancient machinery, and as they cost only half the price of a regular split or solid die perhaps worth investing in for this type of activity. In fact, it is actually *preferable* to use a dienut rather than an ordinary die, as there is less risk of altering the shape and fit of the bolt. However, for general machine-shop use they have little application, and the few that I have are used only on the rare occasions when the moon is seen to be blue!

Cutting action

The cutting action of a die is similar to that for a tap, already discussed, but there is one fundamental difference. *In all but exceptional cases a die must cut a thread full depth.* There is some extrusion effect as in the case of the tap, and this means that there is considerable force tending to expand the die. The result is that a *split* die will, even if adjusted to dead size before use, tend to cut large, and the more use the die has had the greater the effect will be. The die-holder of the normal pattern just has not the strength to resist these forces. The problem is much less with the solid die, but an eventual price has to be paid; once it has become at all blunt the torque necessary to turn the die becomes excessive – in larger sizes it can become unusable. Split dies in the "Pratt & Whitney" type of collet, Fig. 68, are very nearly as good as solid dies but *will* cut large when worn, as the forces are resisted only by small screws in the wall of the collet.

The obvious way of mitigating this problem is to machine the workpiece slightly undersize to allow for the extrusion effect when this is possible, and I follow this practice when necessary, especially when threading stainless steel valve-rods and the like. The exact amount to take off is indeterminate, but I find that between 5% and 10% of the thread height works very well. This may only be 0·002 to 0·004 in. but it does make a great deal of difference. More important is to *keep dies sharp*, and I deal with this matter on page 78.

Unlike the case of the tap, chips can escape quite freely from a die and so long as the die is cutting cleanly it is possible to thread long rods in the lathe at quite high speeds with a suitable lubricant. However, when cutting bright-drawn mild steel, drawn bronze, copper or stainless steel it may be necessary to use the reverse rotation technique to break up the chips. The taper lead-in at the mouth of a die is fairly steep with the deeper threads – say above 6 mm or ¼ in. – and a "used" die some tearing may be experienced with a continuous cut. The answer, again, is to sharpen the die and, perhaps, use a heavier duty cutting oil.

Die-holders

Fig. 69 shows four of different sizes. That at the top is for 1-⁵⁄₁₆ in. dies, next for 1 in., then ¹³⁄₁₆ in. and at the bottom

for the small ⅝ in. size. You can see that there is a larger screw standing at right angles; this is to prevent the die from rotating, though it can also expand the die if the latter is slack in the holder. On either side, not *quite* at 45° to this, is a pair of small grub-screws intended to close the die to its correct size. One of the holders is shown back-to-front so that you can see that there is a shoulder at the bottom of the recess. The die MUST sit flat on this otherwise it will present to the workpiece askew and cut a faulty thread.

A similar fault can be caused by improper use of the centre, locating, screw. If this is tightened down too hard there is a risk of friction at the point displacing one half of the die relative to the other – only with a split die, of course. I have had this happen to such an extent that the die simply stripped the workpiece down to the core diameter! True, it was a fine thread, so that only a small displacement was needed, but I *have* traced malformed threads to this cause in a number of cases. If it happens regularly with one die-holder then it is probable that the point of the

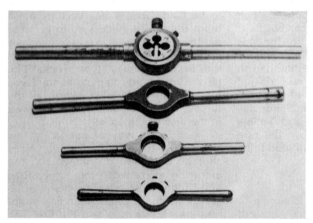

Fig. 69 *A range of die-holders, from 1⁵⁄₁₆ in. (top) down to ⅝ in. dia.*

74

screw is at fault; if with one particular die, then scrap it, for it means that the groove at the split is faulty. This trouble *cannot* occur with a solid die – another of their advantages. My second illustration, Fig. 70, shows the TAILSTOCK DIE-HOLDER, for use in the lathe. At the bottom is a taper shank mandrel which fits the tailstock. Above is the double-ended die-holder, for ¹³⁄₁₆ and 1 in. size dies. There is a hole bored through to fit the mandrel, and it is very important that this be a smooth slide fit – neither tight nor sloppy. These may be bought, but are easy to make – if you need one then you must have a lathe! The machining is quite straightforward except for one point; it is VITAL that the die

recesses be accurately coaxial with the mandrel bore. For years I was plagued with a commercial unit with an error here and poor threads were the norm. It was only a few thousandths of an inch, but it did matter! The best plan is to rough drill the bore, then machine out the die recesses leaving on a machining allowance. You can bore out the centre to a smooth fit on the mandrel (or, better, ream this hole and make the mandrel to suit) and one die recess at a single setting. Then set the body on the normal taper mandrel between centres and, with a tiny boring tool, true up the bore at the other end.

The standard dimensions of dies are as under:

Thread Type	O.D., in.	Thickness, in.	Tolerance on O.D.
All BA*	¹³⁄₁₆	¼	+0/−0·005
Whit. Form			
³⁄₁₆ or less	⅝	¼	+0/−0·005
¹⁄₁₆-¼	¹³⁄₁₆	¼	+0/−0·005
¼-⅜	1	⅜	+0/−0·005
⁵⁄₁₆-⅜	1-⁵⁄₁₆	⁷⁄₁₆	+0/−0·006
⅜-⅝	1½	½	+0/−0·006
½-1	2	⅝	+0/−0·008
I.S.O. Metric			
M1-M2·5	16mm	5mm	+0/−0·12mm
M3-M4	20mm	5mm	+0/−0·12mm
M4-M6	20mm	7mm	+0/−0·12mm
M7-M9	25mm	9mm	+0/−0·12mm
M10-M11	30mm	11mm	+0/−0·15mm
M12-M14	38mm	14mm	+0/−0·15mm
M16-M20	45mm	18mm	+0/−0·15mm
M22-M25	55mm	22mm	+0/−0·2mm

*BA dies below No. 2 may be found ⅝in. dia., and some makers supply Nos. 0, 1 and 2 at 1in. dia.

There is some overlap in "Whitworth Form" (BSW, BSF, M.E. etc.) die diameters and if buying dies individually it is prudent to specify the diameter. Dies sold in sets will use the overlap to reduce the number of sizes to a minimum. I find that a tailstock die-holder for $^{13}\!/_{16}$in. and 1in. dies meets 85% of my requirements.

Lubrication of Taps and Dies

Cast iron needs no lubrication, as the chips are powdery and any lubrication will make an abrasive paste. The graphite in the iron will provide adequate "slipperiness". Almost all other materials will benefit from the use of a lubricant. Unfortunately "the authorities" differ very widely in their recommendations! Most, of course, are concerned with machine tapping, which is not applicable to our work – and in any case the jobbing workshop is unlikely to carry the wide stocks of lubricants found in industrial concerns. I have found the following to be helpful:

Aluminium and alloys	Machine oil and paraffin, 70/30.
Cast brass, GM, bronze	Light machine oil or neat cutting oil.
Ditto, forged, rolled or drawn	Neat cutting oil.
Phos. bronze, drawn	Cutting oil with Rocol additive.
Copper	Machine oil and paraffin, 50/50.
Magnesium alloy	Cutting oil 10%, light machine oil 40%, paraffin 50%
Steel, Mild	Neat cutting oil;
Stainless	Cutting oil with Rocol additive;
Alloy	Neat cutting oil.
Nickel silver	Neat cutting oil.
Silver steel	Neat cutting oil with Rocol additive.

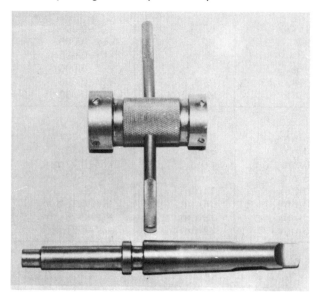

Fig. 70 *A tailstock die-holder for use in the lathe. Double-ended for two sizes of die.*

76

For general tapping I have found that the aerosol spray of Rocol "RTD" oil is very useful, as it can be injected down the hole with little difficulty. The same additive is available from some suppliers as a special tapping and die-cutting oil. It is not cheap, but a little lasts a long time. The main problem with the use of any form of cutting oil on very small taps is that the chips tend to clog easily in the flutes, and it is worth while having a soft brass wire brush with which to clean them after withdrawing the tap for chip clearance. The smaller taps, of course, should not be anointed with heavy oils.

SECTION 10

TAP AND DIE-SHARPENING

Taps can be sharpened both easily and accurately on the Quorn tool and cutter grinder and full instructions are given in the "Quorn Handbook". Those who have no such facilities can, however, tackle the job by hand. The results will not be perfect, but *even an imperfectly sharpened tap is better than a blunt one.*

There are two ways. I use a medium coarse India oilstone – you may have to remove a relatively large amount of metal, and a fine stone is no good for that. On larger taps I use a circular stone about 6 mm (¼ in.) dia., and for smaller ones a tapered round stone which goes down to a point. With these, and lashings of oil, the cutting face in the flute is stoned, keeping the stone well down into the flute so that there is no risk of bevelling the crest of the threads. For small taps – perhaps 6BA and below – even the pointed stone is too large, and for these I use a wedge-shaped stone known as the "Knifeblade", (rather like a gouge-slip but with one edge sharp) India shape No. 28. This can be applied to the leading edges in the tiny grooves. The tap is held in a strong pin-vice which is, in turn, held in the bench vice for these small ones. There is another

shape, No. 53, (the "Machinist's stone") which has a tapered pointed END, and this can be used in the same way. The main problem with tiny taps is to keep the cutting face at a reasonable angle, and I must confess that my own experience shows that after the second sharpening the tap is best thrown away rather than embark on a third attempt.

Note that with almost all taps it will be the first three or four threads that will have worn, and these should be looked at closely with a glass first. You may find that a few strokes of the stone on these alone will mend matters. Finally, DON'T attempt to sharpen a tap off-hand on the bench grinder, nor to convert a broken seconds tap into a plug tap on the same tool! Even a plug tap has a taper lead-in!

Dies are rather easier; all that is needed is a round slipstone – again, not too fine – applied through the clearing holes round the centre threaded portion. But for larger dies a small grinding wheel in a high-speed flexible shaft unit can be used. You must, however, keep careful control of the attitude of the wheel to prevent it from riding up and bevelling the crests. The nearer the wheel is to the hole size the better. I find it necessary to follow with the oilstone and then to use the die on a piece of mild steel stock to remove burrs after grinding. Incidentally, on really large taps you can use the same method, but let the rotation of the wheel be such that this tends to take the wheel *away* from the cutting edges.

You may well find that resharpened taps and dies do not perform quite as well as new ones, for you cannot expect to preserve the carefully formed rake angle which the manufacturer has set up. But you *will* reduce considerably the risk of tap breakage.

Storage of Taps and Dies

The days of the "little tin box" are over! OXO comes in cellophane bags and those who still smoke cigarettes have to make do with thin cardboard. In a way, this is a good thing, for a jumble of taps, drills and dies in a tin box is hardly calculated to preserve the cutting edges. But if tin has gone, various forms of plastic boxes have taken their place, and, fortunately, they can be had very cheaply from watch- and clock-makers sundriesmen in a wide variety of sizes, and, moreover, so designed as to sit neatly on a shelf. However, even a plastic box will not prevent the taps from rubbing their edges against each other or the point of the tapping drill. There is an easy solution for small taps. If the bottom of the box lid is lined with corrugated card, well-dried and then soaked in anti-rust oil (Shell Ensis 254 is my preference, but there are others) then the drill and the taps at least can be separated. Set the taps so that adjacent ones are head-to-tail as an added precaution.

An alternative, using larger boxes of wood or even cardboard, is to cut a

piece of expanded polystyrene packing material to fit the bottom. This can be had free (and with a sigh of relief) from most electrical goods shops, usually in moulded shapes to fit the appliances, but sheet material up to about 20 mm thick is used too. Cut it with a hot knife. Then find a few pieces of steel rod about the same size as the taps and a small "cheese" the same size as the die. Heat this gently and rest it (or them) in the appropriate places on the polystyrene sheet and you will have a series of recesses in which the tools can sit in a few seconds. It just needs a little experiment to find the right temperature, that is all. I use this system for endmills as well as for taps and dies.

I have already shown, in Fig. 37, page 37, my box of tapping drills, home-made, but still reasonable in appearance. Fig. 71 shows a home-made tap, die and drill box made some years ago to house a small set of I.S.O. screwing tackle. The box itself came from a jumble sale and needed new hinges.

Fig. 71 *Home-made case for storing taps, dies, and tapping drills.*

Into this is fixed a thin block of lignum vitae, but any good timber will do — but not mahogany, which, as mentioned, absorbs moisture and will rust everything that touches it. The holes for the dies were excavated with a $^{13}/_{16}$ in. woodworking auger-bit, with finger-grip extensions taken out with a gouge. The slots for the drills and taps were made with a ball-ended slot drill run very fast (about 2400 rpm) in the milling machine but could easily have been done with a small gouge if you have one. The cross-slots at the ends of these tool slots are deeper and enable me to get the tap or drill out simply by pressing down on its end. The lid of the box is filled with oil-soaked foam material — polyurethane, probably — which had been used to pack something that came through the post. It took half an afternoon to make, and is just as good as the boxes supplied, at considerable cost, with commercial sets of screwing tackle.

However, most of my taps and dies are kept in an old (war-time surplus) set of celluloid drawers above the bench, and the contents are prevented from impact either by the use of corrugated cardboard or dividers of some sort. The very small ones (8BA down) are housed in small plastic watchmaker's boxes in these drawers, partly for safety but mainly because it is easier to get hold of them. In which connection a final hint may not be amiss. Cheap plastic tweezers are available from Woolworths and similar places — a godsend when trying to extract a 14BA tap from a small box!

It is worth spending a fair amount of time considering the storage of these tools. It takes very little to chip the edge of a small tap, and these are just the ones which are the most difficult to sharpen and even more difficult to extract if the chip causes them to break in the hole. The manufacturers have taken a lot of care, and spent megabucks on quality control to provide you with first-class tools, and the model you are going to make with them will be equally well cared-for. A pity not to spend a little time on looking after the taps and dies in between?

APPENDIX 'A'

DRILLING FORCES

It has already been remarked that the axial forces involved in the drilling operation are considerably greater than those found when roughing-down in the lathe. There are many published figures but without exception they refer to drilling operations with power down-feed and using "perfectly" formed drill points. Some years ago the author had occasion to make some measurements for another purpose, using hand feed and drills which, while in reasonable condition, were in what might be called "normal workshop state". None were blunt, but for the purposes of the trial no "new" drills were used.

The drilling machine was a Progress No. 1 pillar drill, of 1/2in. capacity fitted with a Jacobs No. 34 chuck. Drive was by vee-belt from a ½hp motor – there was no back gear – and 5-speed pulleys provided 2580, 1325, 1040, 630 and 340 rpm. The measuring device was a substantial set of "bathroom" scales with a maximum capacity of 280 lbf – very solidly made of cast iron and with an iron "platform". The deflection rate was high – about 420 lbf/inch – and the very large dial could be read to intervals of 1 lbf without difficulty.

The test piece was a large (15 lb) slab of mild steel of about 28 tonf/sq.in. UTS, hot rolled and in what appeared to be the normalised state. At each test the drill was brought down and allowed to form a predrill with the corners of the lips about 1/16 in. below the surface. The drill was then fed at a "comfortable" rate by hand and three observations made on each size (in different holes) of the axial force registered on the scale. A final test was made on the larger drills when "hard driving" – feeding as hard as possible. The mean of the results was as under.

Drill dia., in.	Speed rpm	Force lbf
1/8	2580	33
3/16	2580	49
1/4	1325	56
5/16	1325	73
3/8	1040	98
7/16	630	114
1/2	340	134
Hard driving		
3/8	630	155
7/16	340	167
1/2	340	210

The maximum force which could be exerted was off-scale – estimated at 320 lbf.

These figures may be of use as a guide for those interested in building their own bench drilling machines. They also emphasise the advice fre-quently given by the author that, when drilling in the lathe with the drill held in the headstock chuck, with the work on the saddle, the saddle should be pushed forwards using the *tailstock poppet* and NOT by the use either of the rack handwheel or the leadscrew.

APPENDIX B

DIMENSIONS OF BA TAP SHANKS

Several designs of tap holders and guiding devices involve a collet or pin-vice type of grip to the tap. The follow-ing dimensions have been taken from a fairly wide selection of BA taps, includ-ing British, European and American makes.

BA No.	Shank dia., in.	Square A-F, in.
0	0·238/241	0·185/188
1	0·206/210	0·164/166
2	0·192/196	0·145/149
3	0·160/163	0·129/133
4	0·141/143	0·112/114
5 to 12 inc.	0·125/128	0·103/105

A few examples of US manufacture were found to have the uniform dimen-sion of smaller sizes corresponding to the dimensions of the 4BA size rather than the more common 5BA.

APPENDIX C

TABLES

General Note on Tapping Drill Tables

Drill sizes in BOLD type are B.S.I. "Preferred" sizes, but ALL sizes shown in the tables are available. Drills above 14 mm have Morse taper shanks.

The column marked "G.P. Tap Drill" shows the gauge number, letter, or fractional inch size which will give between 60% and 70% thread engagement.

CLEARANCE DRILLS may be ascertained by noting the column giving the thread O.D. in millimetres and adding the desired clearance.

At the head of each table is given a formula which can be used to determine the tapping drill needed to provide ANY desired degree of thread engagement, but see the heading notes in the case of I.S.O. threads. The depth of the male (bolt) thread can be found by subtracting the core diameter from the top diameter and halving the result.

The section on "Tapping drills" on pages 63 et seq., and especially the notes on page 65 should be read before using the tables for the first few times.

TABLE I
INCH/MILLIMETRE CONVERSION
Fractional Inch to Decimal and to Millimetres

inches		mm	inches		mm	inches		mm	inches		mm
	1/64	·01563	0·3969		17/64	·26563	6·7469		33/64	·51563	13·0969		49/64	·76563	19·4469
1/32	3/64	·03125 ·04688	0·7937 1·1906	9/32	19/64	·28125 ·29688	7·1437 7·5406	17/32	35/64	·53125 ·54688	13·4937 13·8906	25/32	51/64	·78125 ·79688	19·8437 20·2406
1/16		·06250	1·5875	5/16		·31250	7·9375	9/16		·56250	14·2875	13/16		·81250	20·6375
	5/64	·07813	1·9844		21/64	·32813	8·3344		37/64	·57813	14·6844		53/64	·82813	21·0344
3/32	7/64	·09375 ·10938	2·3812 2·7781	11/32	23/64	·34375 ·35938	8·7312 9·1281	19/32	39/64	·59375 ·60938	15·0812 15·4781	27/32	55/64	·84375 ·85938	21·4312 21·8281
1/8		·1250	3·1750	3/8		·3750	9·5250	5/8		·6250	15·875	7/8		·8750	22·2250
	9/64	·14063	3·5719		25/64	·39063	9·9219		41/64	·64063	16·2719		57/64	·89063	22·6219
5/32	11/64	·15625 ·17188	3·9687 4·3656	13/32	27/64	·40625 ·42188	10·3187 10·7156	21/32	43/64	·65625 ·67188	16·6687 17·0656	29/32	59/64	·90625 ·92188	23·0187 23·4156
3/16		·18750	4·7625	7/16		·43750	11·1125	11/16		·68750	17·4625	15/16		·93750	23·8125
	13/64	·20313	5·1594		29/64	·45313	11·5094		45/64	·70313	17·8594		61/64	·95313	24·2094
7/32	15/64	·21875 ·23438	5·5562 5·9531	15/32	31/64	·46875 ·48438	11·9062 12·3031	23/32	47/64	·71875 ·73438	18·2562 18·6531	31/32	63/64	·96875 ·98438	24·6062 25·0031
1/4		·250	6·3499	1/2		·500	12·6999	3/4		·750	19·0497	1		1·000	25·400

Decimal inch to mm.

	× 1/1000	× 1/100	× 1/10	INCHES	+10″	+20″
0					254·0	508·0
1	·0254	·254	2·54	25·4	279·4	533·4
2	·0508	·508	5·08	50·8	304·8	558·8
3	·0762	·762	7·62	76·2	330·2	584·2
4	·1016	1·018	10·16	101·6	355·6	609·6
5	·1270	1·270	12·70	127·0	381·0	635·0
6	·1524	1·524	15·24	152·4	406·4	660·4
7	·1778	1·778	17·78	177·8	431·8	685·8
8	·2032	2·032	20·52	203·2	457·2	711·2
9	·2286	2·286	22·86	228·6	482·6	736·6

Millimetres to Inches

	1/1000 mm	1/10 mm	millimetre	+10 mm	+20 mm	+30 mm	+40 mm	+50 mm	+60 mm	+70 mm	+80 mm	+90 mm
0				·39370	·78740	1·1811	1·5748	1·9685	2·3622	2·7559	3·1496	3·5433
1	·00039″	·00394″	·03937″	·44307	·82667	1·2205	1·6142	2·0079	2·4416	2·7953	3·1890	3·5827
2	·00079	·00787	·07874	·47244	·86614	1·2598	1·6535	2·0473	2·4410	2·8347	3·2284	3·6621
3	·00118	·01181	·11811	·51181	·90551	1·2992	1·6929	2·0866	2·4803	2·8740	3·2677	3·6614
4	·00158	·01575	·15748	·55118	·94488	1·3386	1·7323	2·1260	2·5197	2·9134	3·3071	3·7008
5	·00197	·01969	·19685	·59055	·98425	1·3780	1·7717	2·1654	2·5591	2·9258	3·3465	3·7402
6	·00236	·02362	·23622	·62992	1·0236	1·4173	1·8110	2·2047	2·5984	2·9921	3·3858	3·7795
7	·00276	·02756	·27559	·66929	1·0630	1·4567	1·8504	2·2441	2·6378	3·0315	3·4252	3·8189
8	·00315	·03150	·31496	·70866	1·1024	1·4961	1·8898	2·2835	2·6772	3·0709	3·4646	3·8583
9	·00354	·03543	·35433	·74803	1·1417	1·5354	1·9291	2·3228	2·7165	3·1102	3·5039	3·8976

TABLE II
"Preferred sizes" for Parallel Shank Drills

The preferred sizes recommended by the British Standards Institution are intended to encourage designers to limit the number of sizes used on plant and machinery. However, drill *users* often find that non-preferred sizes better suit their needs, especially in the smaller sizes. This applies particularly to tapping drills, where the very small extra cost is more than repaid by the convenience of use. The schedule of "Drill Sizes Available" below is by no means exhaustive, but indicates the range from one reputable manufacturer.

METRIC PREFERRED SIZES
Jobbers Lengths –
flutes from 14 to 8 diameters long.
0·2–0·22–0·25–0·28–0·3mm, and by the same intervals to 1·00mm
1·0mm to 3·0mm by intervals of 0·05mm
3·0mm to 14·0mm by intervals of 0·1mm
14·0mm to 16mm by intervals of 0·25mm
16·0mm to 20mm by intervals of 0·5mm
Stub Lengths –
flutes from 6 to 3 diameters long.
0·5–0·8–1·0–1·2–1·5–1·8–2·0 and by the same intervals to 14mm
14mm to 32mm by intervals of 0·25mm.
"Long" series –
flutes 30 to 10 diameters long.
1·0mm to 14mm by intervals of 0·1mm
14mm to 30mm by intervals of 0·25mm
"Extra Long" –
flutes from 60 to 20 diameters long.
2·0mm to 14mm by intervals of 0·5mm

IMPERIAL PREFERRED SIZES
Jobbers Lengths –
1/64″ to 1/2″ by intervals of 1/64″
Stub Lengths –
1/32″ to 1 1/2″ by intervals of 1/64″
Long Series –
3/64″ to 1 1/4″ by intervals of 1/64″
Extra Long –
1/16″ to 1/2″ by intervals of 1/64″
NOTE: The smaller the drill diameter the greater the relative flute length.

DRILL DIAMETERS AVAILABLE IN JOBBERS LENGTH, PARALLEL SHANK
Sets – 1·0 to 5·0mm; 5·1 to 10mm; 1·0 to 10·0mm, in steps of 0·1mm.
1·0 to 13·0mm in steps of 0·5mm
1/16 to 1/2″ in steps of 1/64″; 1/16″ to 1/2″ in steps of 1/32″
Individual drills – 0·2mm to 2·0mm in 0·01mm steps; 2·0mm to 10·0mm in 0·05mm steps; 10·0mm to 20·0mm in 0·1mm steps, plus intervals at 0·25 and 0·75mm. Direct metric equivalents of Morse number and letter drills. Imperial drills from 1/64″ to 3/4″ by intervals of 1/64″.

Millimetre equivalents of Gauge Number and Letter Size Drills

TABLE III

Gauge No. or Letter	Decimal equivalent (in.)	ALTERNATIVE SIZES B.S.I. Rec. (mm)	ALTERNATIVE SIZES Exact (mm)
80	0.013 5	0.35	0.34
79	0.014 5	0.38	0.37
78	0.016 0	0.40	0.41
77	0.018 0	0.45	0.46
76	0.020 0	0.50	0.51
75	0.021 0	0.52	0.53
74	0.022 5	0.58	0.57
73	0.024 0	0.60	0.61
72	0.025 0	0.65	0.64
71	0.026 0	0.65	0.66
70	0.028 0	0.70	0.71
69	0.029 2	0.75	0.74
68	0.031 0	1/32 in	0.79
67	0.032 0	0.82	0.81
66	0.033 0	0.85	0.84
65	0.035 0	0.90	0.89
64	0.036 0	0.92	0.91
63	0.037 0	0.95	0.94
62	0.038 0	0.98	0.97
61	0.039 0	1.00	0.99
60	0.040 0	1.00	1.02
59	0.041 0	1.05	1.04
58	0.042 0	1.05	1.07
57	0.043 0	1.10	1.09
56	0.046 5	3/64 in	1.18
55	0.052 0	1.30	1.32
54	0.055 0	1.40	1.40
53	0.059 5	1.50	1.51
52	0.063 5	1.60	1.61
51	0.067 0	1.70	1.70
50	0.070 0	1.80	1.78
49	0.073 0	1.85	1.85
48	0.076 0	1.95	1.93
47	0.078 5	2.00	1.99
46	0.081 0	2.05	2.06
45	0.082 0	2.10	2.08
44	0.086 0	2.20	2.18
43	0.089 0	2.25	2.26
42	0.093 5	3/32 in	2.37
41	0.096 0	2.45	2.44
40	0.098 0	2.50	2.49
39	0.099 5	2.55	2.53
38	0.101 5	2.60	2.58
37	0.104 5	2.65	2.64
36	0.106 5	2.70	2.71
35	0.110 0	2.80	2.79
34	0.111 0	2.80	2.82
33	0.113 0	2.85	2.87
32	0.116 0	2.95	2.95
31	0.120 0	3.00	3.05
30	0.128 5	3.30	3.26
29	0.136 0	3.50	3.45
28	0.140 5	9/64 in	3.57
27	0.144 0	3.70	3.66
26	0.147 0	3.70	3.73
25	0.149 5	3.80	3.80
24	0.152 0	3.90	3.86
23	0.154 0	3.90	3.91
22	0.157 0	4.00	3.99
21	0.159 0	4.00	4.04
20	0.161 0	4.10	4.09
19	0.166 0	4.20	4.22
18	0.169 5	4.30	4.30
17	0.173 0	4.40	4.39
16	0.177 0	4.50	4.50
15	0.180 0	4.60	4.57
14	0.182 0	4.60	4.62
13	0.185 0	4.70	4.70
12	0.189 0	4.80	4.80
11	0.191 0	4.90	4.85
10	0.193 5	4.90	4.92
9	0.196 0	5.00	4.98
8	0.199 0	5.10	5.06
7	0.201 0	5.10	5.11
6	0.204 0	5.20	5.18
5	0.205 5	5.20	5.22
4	0.209 0	5.30	5.31
3	0.213 0	5.40	5.41
2	0.221 0	5.60	5.61
1	0.228 0	5.80	5.79
A	0.234 0	5.90	5.94
B	0.238 0	6.00	6.04
C	0.242 0	6.10	6.15
D	0.246 0	6.20	6.25
E	0.250 0	1/4 in	6.35
F	0.257 0	6.50	6.53
G	0.261 0	6.60	6.63
H	0.266 0	17/64 in	6.75
I	0.272 0	6.90	6.90
J	0.277 0	7.00	7.03
K	0.281 0	9/32 in	7.14
L	0.290 0	7.40	7.37
M	0.295 0	7.50	7.49
N	0.302 0	7.70	7.67
O	0.316 0	8.00	8.03
P	0.323 0	8.20	8.20
Q	0.332 0	8.40	8.43
R	0.339 0	8.60	8.61
S	0.348 0	8.80	8.84
T	0.358 0	9.10	9.09
U	0.368 0	9.30	9.34
V	0.377 0	3/8 in	9.58
W	0.386 0	9.80	9.80
X	0.397 0	10.10	10.08
Y	0.404 0	10.30	10.26
Z	0.413 0	10.50	10.49

NOTE

"GAUGE" designation drills are no longer made. In this table "BSI Rec" shows the British Standard "preferred size" drill.

"Exact" is within 0.01mm (0.00039") of the old gauge size. "Preferred" drills are cheaper.

TABLE IV   COMBINATION CENTRE (SLOCOMBE) DRILLS

Current Standard types   See Fig. Below

All Dimensions in mm.

Pilot Dia. "d" *Ins.* mm *(approx.)*	Body Dia. D mm "A"	"B"	"C"	Guard Dia. "d" mm	Body Length (Max.) mm "A"	"B"	"C"	Pilot Length (Max.) mm "A"/"B"	"C"	Point Rad. (Max.) mm	(Min.) mm
³⁄₆₄ 1·0	3·15	4·0	3·15	2·12	33	37	33	1·9	3·0	33·5	29·5
¹⁄₁₆ 1·6	4·0	6·3	4·0	3·35	37	47	37	2·8	4·25	37·5	33·5
⁵⁄₆₄ 2·0	5·0	8·0	5·0	4·25	42	52	42	3·3	5·4	42	38
³⁄₃₂ 2·5	6·3	10·0	6·3	5·30	47	59	47	4·1	6·7	47	43
¹⁄₈ 3·15	8·0	11·2	8·0	5·70	52	63	52	4·9	8·5	52	48
⁵⁄₃₂ 4·0	10·0	14·0	10·0	8·50	59	70	59	6·2	10·6	59	53
¼ 6·3	16·0	20·0	16·0	13·20	74	83	74	9·2	17·0	74	68

Old (1950) Standard
Type "A", but point at 118°

All Dimensions in inches

No.	"d"	"D"	"L" Max.	"I" Max.
BS1	³⁄₆₄	⅛	1½	⁵⁄₆₄
BS2	¹⁄₁₆	³⁄₁₆	1¾	³⁄₃₂
BS3	³⁄₃₂	¼	2	⁵⁄₃₂
BS4	⅛	⁵⁄₁₆	2¼	³⁄₁₆
BS5	³⁄₁₆	⁷⁄₁₆	2½	⁹⁄₃₂
BS6	¼	⅝	3¹¹⁄₃₂	¹¹⁄₃₂

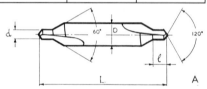

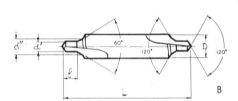

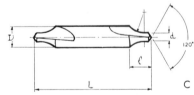

The "Slocombe" or Combination centre drill. (a) The normal type. (b) Type producing a guard recess. (c) Precision type producing a curved profile to the centre-hole.

TABLE V   TAPPING DRILLS

British Standard Whitworth Thread

Tapping drill dia.=D − (1·28P×E/100) inches

Dia. Ins	TPI	Core dia. ins	Core Area sq. ins	50%	55%	60%	65%	70%	75%	G.P. Tap drill	Thread O.D. mm
				Tapping Drill Dia. − mm							
1/16	60	0·041	0·0013	1·32	1·29	1·25	1·23	1·22	1·18	1·25mm	1·587
3/32	48	0·067	0·0035	2·05	2·0	1·98	1·95	1.92	1·88	5/64	2·381
1/8	40	0·093	0·0068	2·78	2.74	2·7	2·65	2·60	2·58	No. 37	3·175
5/32	32	0·116	0·0106	3·45	3·4	3·35	3·30	3.25	3·2	No. 30	3·969
3/16	24	0·134	0·014	4·09	4·0	3·95	3·86	3·8	3·75	No. 24	4·763
1/4	20	0·186	0·027	5·55	5·45	5·35	5·30	5·22	5·13	No. 4	6·35
5/16	18	0·241	0·046	7·05	6·95	6·85	6·75	6·65	6·6	H	7·938
3/8	16	0·295	0·068	8·5	8·4	8·3	8·2	8·1	8·0	P	9·525
7/16	14	0·346	0·094	9·9	9·8	9·7	9·6	9·5	9·4	V	11·113
1/2	12	0·393	0·121	11·3	11·2	11·1	11·0	10·8	10·6	7/16	12·7
5/8	11	0·509	0·203	14·4	14·25	14·1	13·9	13·8	13·7	35/64	15·875
3/4	10	0·622	0·304	17·4	17·25	17·1	16·9	16·75	16·6	21/32	19·05
7/8	9	0·733	0·422	20·5	20·25	20·0	19·84	19·75	19·5	25/32	22·225
1	8	0·840	0·554	—	23·2	23·0	22·75	22·5	22·3	29/32	25·4

BRITISH STANDARD FINE THREAD (Whitworth form)

Tapping Drill Dia.=D − (1·28P×E/100) inches

Dia. Ins	TPI	Core dia. ins	Core Area sq. ins	50%	55%	60%	65%	70%	75%	G.P. Tap drill	Thread O.D. mm
				Tapping Drill Dia. − mm							
3/16	32	0·148	0·017	4·25	4·2	4·15	4·1	4·05	4·0	No. 20	4·763
1/4	26	0·201	0·032	5·72	5·66	5·6	5·54	5·47	5·4	7/32	6·35
5/16	22	0·254	0·051	7·2	7·13	7·05	6·98	6·9	6·83	I	7·938
3/8	20	0·311	0·076	8·72	8·64	8·55	8·47	8·4	8·3	Q	9·525
7/16	18	0·366	0·105	10·2	10·1	10·0	9·9	9·85	9·76	25/64	11·113
1/2	16	0·420	0·138	11·7	11·6	11·5	11·4	11·3	11·2	29/64	12·7
5/8	14	0·483	0·224	14·7	14·6	14·5	14·4	14·25	14·1	9/16	15·875
3/4	12	0·643	0·325	17·7	17·6	17·4	17·3	17·1	17·0	43/64	19·05
7/8	11	0·759	0·452	20·75	20·6	20·4	20·3	20·2	20·0	51/64	22·225
1	10	0·872	0·597	23·75	23·6	23·5	23·3	23·1	22·9	29/32	25·4

The General Purpose (GP) Tapping drills give between 60 and 70% thread engagement in this table.

TABLE VI   TAPPING DRILLS

Model Engineer and Special Threads, Whitworth Form

Tapping drill dia.$=D-(1.28P\times E/100)$ inches

Dia. Ins	Dia. mm	Core dia. Ins	Core Area Sq. ins.	Tapping Drill Dia. – mm						G.P. Tap drill No./ins
				60%	65%	70%	75%	80%	85%	
40 TPI Series Thread Height 0·016″=0·406mm										
⅛	3·175	0·093	0·0068	2·70	2·65	2·60	2·58	2·55	2·48	No. 37
⁵⁄₃₂	3·97	0·125	0·0123	3·5	3·45	3·40	3·35	3·30	3·28	No. 29
³⁄₁₆	4·763	0·155	0·0189	4·3	4·22	4·20	4·15	4·10	4·07	No. 19
⁷⁄₃₂	5·556	0·187	0·0275	5·10	5·05	5·0	4·95	4·90	4·86	No. 9
¼	6·35	0·218	0·0373	5·85	5·80	5·78	5·75	5·70	5·66	I
⁹⁄₃₂	7·144	0·249	0·0487	6·65	6·63	6·60	6·55	6·50	6·45	G
⁵⁄₁₆	7·938	0·282	0·063	7·45	7·4	7·37	7·35	7·30	7·25	L
⅜	9·525	0·343	0·0924	9·05	9·0	0·95	8·90	8·85	8·83	S
32 TPI Series Thread Height 0·020″=0·508mm										
¼	6·35	0·210	0·0346	5·75	5·70	5·65	5·60	5·55	5·50	No. 2
⁹⁄₃₂	7·144	0·241	0·0456	6·55	6·50	6·45	6·40	6·35	6·30	F
⁵⁄₁₆	7·038	0·272	0·0581	7·35	7·30	7·25	7·20	7·15	7·10	K
⅜	9·525	0·335	0·088	8·90	8·85	8·80	8·75	8·70	8·65	S
⁷⁄₁₆	11·11	0·398	0·124	10·5	—	10·4	—	10·3	10·25	Z
½	12·7	0·460	0·166	12·1	—	12·0	—	11·9	11·8(88%)	¹⁵⁄₃₂″
26 TPI Series Thread Height 0·0246″=0·625mm (Brass Gas thread)										
³⁄₁₆	4·7634	0·138	0·015	4·0	3·95	3·89	3·82	3·76	3·70	No. 23
¼	6·35	0·201	0·032	5·60	5·54	5·47	5·41	5·35	5·29	⁷⁄₃₂
⁵⁄₁₆	7·938	0·263	0·054	7·19	7·12	7·06	7·0	6·94	6·87	K
⅜	9·525	0·326	0·083	8·77	8·71	8·65	8·59	8·52	8·46	¹¹⁄₃₂
⁷⁄₁₆	11·11	0·388	0·12	10·36	10·30	10·23	10·17	10·11	10·05	Y
½	12·7	0·451	0·16	11·95	11·89	11·82	11·76	11·70	11·64	¹⁵⁄₃₂
⅝	15·88	0·576	0·26	15·13	15·06	15·0	14·94	14·88	14·82	¹⁹⁄₃₂
¾	19·05	0·701	0·39	18·30	18·24	18·17	18·11	18·05	18·0	²³⁄₃₂
60 TPI Series Thread Height 0·0107″=0·271mm										
¹⁄₁₆	1·588	0·041	0·0013	1·26	1·23	1·21	1·18	1·15	N.R.	1·25mm
³⁄₃₂	2·38	0·083	0·0054	2·06	—	2·0	1·97	1·96	N.R.	No. 47
⅛	3·175	0·114	0·0103	2·85	2·82	2·80	2·75	—	2·71	No. 34
⁵⁄₃₂	3·967	0·145	0·0165	3·65	3·60	—	3·57	3·55	3·50	No. 28
³⁄₁₆	4·763	0·177	0·0246	4·45	4·40	4·39	4·35	—	4·30	No. 17

N.R.=Not Recommended at the % engagement unless exceptional circumstances demand it.

TABLE VII TAPPING DRILLS

British Association (BA) Threads

Tapping drill dia.=D – (1·2P×E/100) inches

BA No.	O.D. mm	Pitch mm	TPI Approx. Exact	Core Dia. mm	Core Area mm²	Tapping Drill Dia. – mm 60%	65%	70%	75%	G.P. Drill No.	BA No.
0	6·0	1·0	25·4	4·8	18·1	5·3	5·22	5·16	**5·1**	No. 5	0
1	5·3	0·9	28·25	4·22	14·0	4·65	**4·6**	4·55	**4·50**	No. 14	1
2	4·7	0·81	31·35	3·73	10·9	4·1	**4·0**	3·95	**3·9**	No. 22	2
3	4·1	0·73	34·84	3·22	8·14	3·57	**3·5**	3·45	**3·4**	No. 29	3
4	3·6	0·66	38·46	2·81	6·20	3·15	**3·1**	3·05	**3·0**	No. 31	4
5	3·2	0·59	43·10	2·49	4·87	2·78	**2·75**	**2·7**	**2·65**	No. 36	5
6	2·8	0·53	47·85	2·16	3·67	2·44	**2·4**	2·38	**2·35**	No. 41	6
7	2·5	0·48	52·91	1·92	2·89	**2·15**	2·1	2·08	**2·05**	No. 45	7
8	2·2	0·43	59·17	1·68	2·22	**1·89**	1·86	1·84	**1·8**	No. 49	8
9	1·9	0·39	64·94	1·43	1·61	1·62	**1·6**	1·57	**1·55**	No. 52	9
10	1·7	0·35	72·46	1·28	1·29	**1·45**	**1·42**	**1·4**	1·38	No. 54	10
11	1·5	0·31	81·97	1·13	1·00	1·28	1·26	1·24	**1·20**	³⁄₆₄"	11
12	1·3	0·28	90·91	0·96	0·724	**1·10**	1·08	1·06	**1·05**	No. 57	12
13	1·2	0·25	102·04	0·90	0·636	1·02	**1·00**	0·99	0·97	No. 60	13
14	1·0	0·23	109·9	0·72	0·407	0·83	**0·82**	0·81	0·79	No. 66	14
15	0·9	0·21	120·5	0·65	0·332	**0·75**	0·74	**0·72**	0·71	No. 69	15
16	0·79	0·19	133·3	0·56	0·246	**0·65**	0·64	0·63	**0·62**	No. 72	16
17	0·70	0·17	149·4	0·50	0·196	**0·58**	0·57	0·56	**0·55**	No. 74	17
18	0·62	0·15	169·3	0·44	0·150	0·51	**0·50**	0·49	**0·48**	No. 76	18

TABLE VIII   TAPPING DRILLS
I.S.O. METRIC THREADS
'M' Designation to B.S.3643 and 'S' Designation to B.S. 4827

IMPORTANT: See notes on page 50

Tapping drill dia. for "M" designation=D − (1·083P×E/100): For "S" designation=D − (0·096P×E/100) where E=% Flank height.

Column headings of "%" based on the flank contact. [] indicates % of nut thread depth, and () is % of bolt thread depth.

Size mm	Pitch mm	Core Dia. mm	Core Area mm²	60% [63] (53)	65% [68] (57)	70% [73] (62)	75% [77] (66)	80% [82] (70·5)	85% [86] (75)	B.A. Equivalent
S 0·5	0·125	0·369	0·107	0·43	**0·42**	—	0·41	**0·40**		
S 0·6	0·150	0·432	0·147	0·51	—	**0·50**	0·49	**0·48**	Not	18
S 0·7	0·175	0·504	0·199	**0·6**	0·59	**0·58**	0·57	**0·56**	Recom-	17
S 0·8	0·20	0·576	0·261	0·69	**0·68**	0·67	0·66	0·65	mended	16
S 1·0	0·25	0·720	0·407	**0·85**	0·84	0·83	**0·82**	0·81		14
S 1·2	0·25	0·920	0·664	**1·04**	1·04	1·03	1·02	1·01		13
M 1·0	0·25	0·694	0·378	0·84	0·82	0·81	**0·80**	0·78		14
M 1·2	0·25	0·894	0·627	1·04	1·02	1·01	0·99	0·98	Not	13
M 1·4	0·30	1·033	0·838	1·21	1·19	1·18	1·16	1·14	Recom-	11/12
M 1·6	0·35	1·171	1·007	1·37	**1·35**	1·33	1·32	**1·30**	mended	10/11
M 1·8	0·35	1·371	1·476	1·57	**1·55**	1·53	1·52	**1·5**		9/10
M 2·0	0·40	1·51	1·79	1·74	1·72	**1·70**	1·67	**1·65**	1·63	9
M 2·2	0·45	1·65	2·14	**1·9**	1·88	1·86	1·84	1·81	1·79	8
M 2·5	0·45	1·95	2·99	**2·20**	2·18	**2·15**	—	**2·10**	2·09	7
M 3·0	0·50	2·39	4·49	**2·70**	2·65	2·62	**2·60**	2·58	2·55	5/6
M 3·5	0·60	2·77	6·00	**3·10**	3·05 ←→ 3·05		**3·00**	2·95 ←→ 2·95		4
M 4·0	0·70	3·14	7·74	3·55	**3·50**	3·45 ←→ 3·45		**3·40**	3·35	3
M 4·5	0·75	3·58	10·06	**4·00**	3·97	3·95	**3·90**	3·85	3·81	3/2
M 5·0	0·80	4·02	12·7	**4·50**	4·45	**4·40**	4·35	**4·30**	4·25	2/1
M 6·0	1·00	4·78	17·9	5·35	**5·30**	5·25	**5·20**	5·15	**5·10**	0
M 8·0	1·25	6·47	32·9	**7·20**	**7·10**	7·05	6·95	**6·90**	6·85	
M 10	1·50	8·15	52·2	**9·2**	**9·0**	**8·9**	**8·8**	**8·7**	**8·6**	
M 12	1·75	9·86	76·4	**10·9**	**10·8**	**10·7**	**10·6**	**10·5**	**10·4**	
M 16	2·0	13·55	144·2	14·7	14·6	14·5	14·4	14·3	14·2	
M 20	2·5	16·95	225·6	18·4	18·2	18·1	18·0	17·8	17·7	
M 24	3·0	20·33	324·6	22·0	21·9	21·7	21·6	21·4	21·25	

Note: Drills over 12mm dia. are taken from the taper shank list, though straight shank drills may be available up to 20mm dia.

←→ indicates that drill shown lies midway between the two columns.

BOLD figures are British Standard "Preferred sizes".

92

TABLE IX   TAPPING DRILLS

I.S.O. Metric Threads – Fine Pitch Series, To B.S.3643. See Notes on page 50 and heading of Table VIII.

ISO fine threads should be designated by the prefix 'M' and the pitch, thus – M2·0×2.5.

Size mm	Pitch mm	Core Dia. mm	Core Area mm²	Tapping Drill Dia. – mm					Approx. BA Equiv.
				60% [63] (53)	65% [68] (57)	70% [73] (62)	75% [77] (66)	80% [82](70·5)	
1·0	0·2	0·755	0·52	0·87	0·86	**0·85**	0·84	0·83	14
1·1	0·2	0·855	0·65	0·97	0·96	**0·95**	0·94	0·93	
1·2	0·2	0·955	0·81	1·07	1·06	**1·05**	1·04	1·03	13
1·4	0·2	1·155	1·16	1·27	1·26	**1·25**	1·24	1·23	11/12
1·6	0·2	1·355	1·57	1·47	1·46	**1·45**	1·44	1·43	10/11
1·8	0·2	1·555	2·04	1·67	1·66	**1·65**	1·64	1·63	9/10
2·0	0·25	1·69	1·84	1·84	1·83	1·81	**1·80**	1·78	9
2·2	0·25	1·89	3·03	**2·05**	—	**2·0** ←→	**2·0**	1·98	8
2·5	0·35	2·07	3·71	—	2·26	**2·25**	—	**2·20**	7
3·0	0·35	2·57	5·61	2·78	**2·75** ←→	**2·75**	2·71	**2·70**	5/6
3·5	0·35	3·07	7·90	3·26	**3·25**	—	3·23	**3·20**	4
4·0	0·5	3·39	9·8	**3·70**	3·65	—	**3·6**	3·55	3
4·5	0·5	3·89	12·8	**4·20**	4·15	—	**4·10**	4·05	3/2
5·0	0·5	4·39	16·1	**4·70**	4·65	4·62	**4·60**	4·57	2/1
6·0	0·75	5·08	22·0	**5·50**	5·48	5·43	**5·40**	5·35	0
8·0	1·0	6·08	31·3	7·35	**7·30**	7·25	**7·20**	7·15	

Column figures of "%" based on flank contact. [] indicates % of tap engagement and () is % of bolt thread depth.

←→ indicates that drill shown lies midway between the two columns.

BOLD figures are British Standard "Preferred sizes".

TABLE X TAPPING DRILLS

Metric Constant Pitch Threads for Model Engineers, to BS.PD6507-1982

IMPORTANT: See Notes on page 53

Tapping drill diameter, based on depth of flank engagement, $=D - (1.083P \times E/100)$

Size mm	Pitch mm	Core Dia. mm	Core Area mm²	Tapping Drill Dia. – mm				
				60% [63]	65% [68]	70% [73]	75% [77]	80% [82]
3·0	0·5	2·38	4·45	—	2·65	—	2·60	2·58
4·0	0·5	3·39	9·02	—	3·65	—	3·60	3·57
4·5	0·5	3·90	11·95	4·20	4·15	—	4·10	4·05
5·0	0·5	4·39	15·10	4·70	4·65	4·62	4·60	4·57
5·5	0·5	4·90	18·9	—	5·15	—	5·10	5·06
6·0	0·5	5·39	22·8	—	5·65	5·63	5·60	5·56
4·5	0·75	3·58	10·10	4·0	3·97	3·95	3·9	3·86
6·0	0·75	5·08	20·2	5·50	5·48	5·43	5·40	5·35
7·0	0·75	6·08	29·0	6·53	6·50	6·45	6·40	6·35
8·0	0·75	7·08	39·4	—	7·5	7·45	7·40	7·35
10·0	0·75	9·08	64·7	—	9·50	9·45	9·40	9·35
12·0	0·75	11·08	96·4	11·51	11·50	—	11·40	11·30**
10·0	1·0	8·77	60·4	9·35	9·30	9·25	9·20	9·15
12·0	1·0	10·77	91·1	11·35	11·30	11·25	11·20	11·1**
14·0	1·0	12·77	128·0	13·4	13·3	13·25	13·20	13·1**
16·0	1·0	14·77	171	15·4	15·3	15·25	15·2	15·1**
18·0	1·0	16·77	221	17·4	17·3	17·25	17·2	17·1**
20·0	1·0	18·77	277	19·4	19·3	19·25	19·2	19·1**

Figures in brackets [] indicate nut thread % engagement – i.e. the depth "seen" by the tap.
** These drills will provide about 84% flank engagement.

TABLE XI TAPPING DRILLS

Unified Coarse (UNC) and Fine (UNF) Threads

Tapping Drill Dia.=D − (1·082P×E/100) based on flank engagement.

(With rounded root to nut thread the tap engagement will be about 3% greater)

U.N.C. Dia. ins.	Core Dia. ins.	Core Area ins²	T.P.I.	Tapping Drill Dia. − mm					Thread O.D. mm
				55%	60%	65%	70%	75%	
¼	0·187	0·028	20	**5·60**	**5·50**	5·45	**5·4**	5·31*	6·35
⁵⁄₁₆	0·243	0·046	18	**7·1**	**7·00**	6·95	**6·90**	**6·80**	7·94
⅜	0·297	0·069	16	**8·60**	**8·50**	**8·40**	8·35	8·25	9·53
⁷⁄₁₆	0·394	0·096	14	**10·0**	9·92	9·85	9·75	9·65	11·11
½	0·404	0·128	13	**11·5**	**11·4**	**11·3**	11·25	11·11*	12·70
⅝	0·512	0·206	11	14·5	14·4	14·25	14·1	14·0	15·88
¾	0·625	0·307	10	17·5	17·4	17·25	17·1	17·0	19·05
⅞	0·737	0·427	9	20·5	20·4	20·25	20·1	19·9	22·23
1	0·847	0·563	8	23·5	23·3	23·2	23·0	22·8	24·40
U.N.F.									
¼	0·205	0·033	28	**5·80**	5·75	**5·70**	5·65	**5·60**	6·35
⁵⁄₁₆	0·260	0·053	24	**7·30**	7·25	**7·20**	7·15	**7·10**	7·94
⅜	0·323	0·082	24	**8·90**	8·85	8·75	8·73*	8·65	9·53
⁷⁄₁₆	0·375	0·110	20	**10·30←→10·3**	10·25	**10·20**	10·08	11·11
½	0·438	0·151	20	**11·90←→11·90**	**11·80**	11·75	**11·60**	12·70
⅝	0·556	0·243	18	15·08*	15·0	14·9	14·8	14·75	15·88
¾	0·672	0·355	16	18·1	18·0	17·9	17·86*	17·75	19·05
⅞	0·876	0·485	14	21·1	21·0	21·0	20·9	20·75	22·23
1	0·896	0·631	12	—	24·0 ←→24·0		23·81	23·5	25·40

Note: * indicates a Morse number or fractional Imperial equivalent drill.

TABLE XII   TAPPING DRILLS

Small U.S. National Fine and Coarse (N.F. and N.C.) Threads

Tapping Drill Dia.taken as D – (1·3P×E/100). See page 50. Bolt thread height 0·614P.

# No.	Dia. ins.	T.P.I.	Core Dia. ins.	Core Area ins²	Tapping Drill Dia. – mm					Thread O.D. mm
					55%	60%	65%	70%	75%	
0–80	0·060	80	0·0447	0·0016	1·29	1·27	**1·25**	1·23	1·21	1·524
1–64	0·073	64	0·0538	0·0023	1·57	**1·55**	1·52	1·49	1·47	1·854
1–72	0·073	72	0·0560	0·0025	**1·60**	1·58	**1·55**←→**1·55**		1·51*	1·854
2–56	0·086	56	0·0641	0·0032	1·86	1·83	**1·80**	1·77	1·75	2·184
2–64	0·086	64	0·0668	0·0035	**1·90**	1·88	**1·85***	1·82	**1·80**	2·184
3–48	0·099	48	0·0734	0·0042	**2·15**	**2·10**	2·07	**2·05**	**2·00**	2·515
3–56	0·099	56	0·0771	0·0047	**2·20**	2·18	**2·15**	**2·10**	2·07	2·515
4–40	0·112	40	0·0813	0·0052	**2·40**	**2·35**	**2·30**	2·26*	**2·20**	2·845
4–48	0·112	48	0·0864	0·0059	**2·45**	2·44*	**2·40**	**2·35**	**2·30**	2·845
5–40	0·125	40	0·0943	0·007	**2·70**	**2·65**←→2·64*	**2·60**	**2·55**		3·175
5–44	0·125	44	0·0971	0·0074	**2·75**	2·71*	**2·70**	**2·65**	**2·60**	3·175
6–40	0·138	40	0·1073	0·009	3·05*	**3·0**	2·95*	**2·90**	**2·85**	3·505
8–32	0·164	32	0·1257	0·0124	**3·60**	3·55	**3·50**	3·45*	**3·40**	4·166
8–36	0·164	36	0·1299	0·0135	3·66*	**3·60**	3·57*	**3·50**←→**3·50**		4·166
10–24	0·190	24	0·1389	0·0152	4·05	**4·0**	3·91*	3·86*	**3·80**	4·826
10–32	0·190	32	0·1517	0·0181	4·25	**4·20**	4·15	**4·10**	4·05	4·826
12–24	0·216	24	0·1649	0·0214	4·70*	4·65	**4·60**	4·55	4·45	5·486
12–28	0·216	28	0·1772	0·0233	4·85*	4·76	4·70*	4·65	**4·60**	5·486

Drills marked * are an exact equivalent to Morse "Number" drills.

TABLE XIII   TAPPING DRILLS

British Standard Pipe Threads (Parallel) Whitworth form.

Tapping drill diameter$=$D $-$ (1·28P$\times$E/100)

Nom. Dia.	O.D. ins	TPI	Core Dia. ins	Core Area ins^2	Tapping Drill Dia. – mm or inch				G.P. Tap drill	Thread O.D. mm
					70%	75%	80%	85%		
⅛	0·383	28	0·337	0·089	8·9	8·85	8·80	8·75	S	9·73
¼	0·518	19	0·451	0·160	12·0	11·9	11·8	11·7	¹⁵/₃₂″	13·16
⅜	0·656	19	0·589	0·273	15·5	³⁹/₆₄″	15·4	15·3	³⁹/₆₄″	16·66
½	0·825	14	0·734	0·423	19·4	19·25	19·1	¾″	¾″	20·96
¾	1·041	14	0·950	0·709	24·75	³¹/₃₂″	24·5	24·3	³¹/₃₂″	26·44

TABLE XIV

Sparking Plug Threads – SAE Standard. 60° Thread Form.

Tapping drill diameter=D – (1·25P×E/100)

Dia. mm	Pitch mm	Core Dia. mm	Core Area mm^2	Tapping Drill Dia. – mm		
				60%	70%	80%
10	1·0	8·75	60·1	9·15	9·10	9·0
12	1·25	10·44	85·6	11·10	10·9	10·75
14	1·25	12·44	121·5	13·10	12·9	12·75
18	1·5	15·75	194·8	16·8	16·7	16·5

TABLE XV   INDEX OF TAPPING DRILLS

This table is "drill-based" and shows the threads and % thread engagement which is provided by the classical metric drill set of 1mm to 10mm×0·1mm steps.

NOTE Unless otherwise stated threads in the "Model Engineer" (M.E.) column are all 40 tpi.

Drill	BA	**M.E.**	BSW	BSF	ISO	UNC/UNF
1·0	13 65%				M1·2 73%	
1·1	12 70%					
1·2	11 75%	¹⁄₁₆–60 72%	¹⁄₁₆ 72%		M1·4 63%	0–80 77%
1·3	11 55%		¹⁄₁₆ 54%		M1·6 80%	
1·4	10 70%					1–64 68%
1·5					M1·8 80%	1–72 76%
1·6	9 65%					1–72 55%
1·7					M2·0 70%	2–64 75%
1·8	8 75%				M2·2 83%	2-56 65%
1·9	8 58%		³⁄₃₂ 73%		M2·2 62%	
2·0	7 82%	³⁄₃₂–60 70%	³⁄₃₂ 55%			
2·1	7 65%				M2·5 80%	3–56 70%
2·2					M2·5 60%	4–40 75%
2·3						4–40 65%
2·4	6 65%					4–48 65%
2·5						
2·6	5 80%	¹⁄₈ 70%	¹⁄₈ 70%		M3·0 75%	5–40 70%
2·7	5 70%	¹⁄₈ 60%	¹⁄₈ 60%		M3·0 60%	5–44 65%
2·8	5 59%	¹⁄₈–60 70%				
2·9		¹⁄₈–60 52%				6–40 70%
3·0	4 75%				M3·5 75%	6–40 60%
3·1	4 65%				M3·5 60%	
3·2			⁵⁄₃₂ 75%			
3·3			⁵⁄₃₂ 65%			
3·4	3 75%	⁵⁄₃₂ 70%	⁵⁄₃₂ 55%		M4·0 80%	8–32 75%
3·5	3 65%	⁵⁄₃₂ 60%			M4·0 65%	8–32 65%
3·6	3 57%	⁵⁄₃₂–60 65%				8–36 60%
3·7						
3·8		³⁄₁₆–26 77%	³⁄₁₆ 70%			10–24 75%
3·9	2 75%	³⁄₁₆–26 70%	³⁄₁₆ 62%		M4·5 75%	10–24 66%
4·0	2 65%	³⁄₁₆–26 60%		³⁄₁₅ 75%	M4·5 60%	10–24 60%

Drill	BA	**M.E.**	BSW	BSF	ISO	UNC/UNF
4·1		³⁄₁₆ 80%		³⁄₁₆ 65%		10–32 70%
4·2		³⁄₁₆ 70%		³⁄₁₆ 55%		10–32 60%
4·3		³⁄₁₆ 60%			M5·0 80%	
4·4		³⁄₁₆–60 65%			M5·0 71%	
4·5	1 75%				M5·0 60%	12–24 73%
4·6	1 65%					12–24 65%
4·7	1 55%					12–28 65%
4·8						12–28 57%
4·9		⁷⁄₃₂ 80%				
5·0		⁷⁄₃₂ 70%				
5·1	0 75%	⁷⁄₃₂ 60%			M6 75%	
5·2	0 67%		¼ 71%		M6 65%	
5·3	0 60%		¼ 65%			
5·4		¼–26 75%	¼ 57%	¼ 75%		¼UNC 70%
5·5		¼–26 67%	¼ 53%	¼ 67%		¼UNC 60%
5·6		¼–32 75%		¼ 60%		¼UNF 75%
5·7		¼–32 65%		¼ 52%		¼UNF 65%
5·8		¼–40 65%				¼UNF 55%
5·9						
6·0						
6·1						
6·2						
6·3						
6·4		⁹⁄₃₂–32 75%				
6·5		⁹⁄₃₂–32 65%				
6·6		⁹⁄₃₂ 70%	⁵⁄₁₆ 75%			
6·7			⁵⁄₁₆ 67%			
6·8			⁵⁄₁₆ 58%			⁵⁄₁₆UNC 75%
6·9			⁵⁄₁₆ 58%	⁵⁄₁₆ 70%	M8 80%	⁵⁄₁₆UNC 70%
7·0		⁵⁄₁₆–26 75%	⁵⁄₁₆ 50%	⁵⁄₁₆ 60%	M8 70%	⁵⁄₁₆UNC 60%
7·1		⁵⁄₁₆–26 67%		⁵⁄₁₆ 52%	M8 65%	⁵⁄₁₆UNF 75%
7·2		⁵⁄₁₆–32 75%				⁵⁄₁₆UNF 65%
7·3		⁵⁄₁₆–32 65%				⁵⁄₁₆UNF 55%
7·4		⁵⁄₁₆–40 65%				
7·5						
7·6						
7·7						
7·8						
7·9						
8·0			³⁄₈ 75%			

Drill	BA	**M.E.**	BSW	BSF	ISO	UNC/UNF
8·1			3/8 70%			
8·2			3/8 65%			
8·3			3/8 60%	3/8 75%		3/8UNC 73%
8·4			3/8 55%	3/8 70%		3/8UNC 65%
8·5				3/8 63%		3/8UNC 60%
8·6		3/8–26 74%		3/8 57%		
8·7		3/8–26 64%			M10 80%	3/8UNF 72%
8·8		3/8–32 70%			M10 75%	3/8UNF 63%
8·9		3/8–40 75%			M10 70%	3/8UNF 55%
9·0		3/8–40 65%			M10 65%	
9·1						
9·2						
9·3						
9·4			7/16 75%			
9·5			7/16 70%			
9·6			7/16 65%			
9·7			7/16 60%			7/16UNC 73%
9·8			7/16 55%	7/16 72%		7/16UNC 68%
9·9				7/16 65%		7/16UNC 62%
10·0				7/16 60%		7/16UNF 80%

Index